LEAD-FREE SOLDERING IN ELECTRONICS

LEAD-FREE SOLDERING IN ELECTRONICS

SCIENCE, TECHNOLOGY, AND ENVIRONMENTAL IMPACT

EDITED BY

KATSUAKI SUGANUMA

Osaka University
Osaka, Japan

MARCEL DEKKER, INC.   NEW YORK · BASEL

Although great care has been taken to provide accurate and current information, neither the author(s) nor the publisher, nor anyone else associated with this publication, shall be liable for any loss, damage, or liability directly or indirectly caused or alleged to be caused by this book. The material contained herein is not intended to provide specific advice or recommendations for any specific situation.

Trademark notice: Product or corporate names may be trademarks or registered trademarks and are used only for identification and explanation without intent to infringe.

Library of Congress Cataloging-in-Publication Data
A catalog record for this book is available from the Library of Congress.

ISBN: 0-8247-4102-1

This book is printed on acid-free paper.

Headquarters
Marcel Dekker, Inc., 270 Madison Avenue, New York, NY 10016, U.S.A.
tel: 212-696-9000; fax: 212-685-4540

Distribution and Customer Service
Marcel Dekker, Inc., Cimarron Road, Monticello, New York 12701, U.S.A.
tel: 800-228-1160; fax: 845-796-1772

Eastern Hemisphere Distribution
Marcel Dekker AG, Hutgasse 4, Postfach 812, CH-4001 Basel, Switzerland
tel: 41-61-260-6300; fax: 41-61-260-6333

World Wide Web
http://www.dekker.com

The publisher offers discounts on this book when ordered in bulk quantities. For more information, write to Special Sales/Professional Marketing at the headquarters address above.

Copyright © 2004 by Marcel Dekker, Inc. All Rights Reserved.

Neither this book nor any part may be reproduced or transmitted in any form or by any means, electronic or mechanical, including photocopying, microfilming, and recording, or by any information storage and retrieval system, without permission in writing from the publisher.

Current printing (last digit):

10 9 8 7 6 5 4 3 2 1

PRINTED IN THE UNITED STATES OF AMERICA

Foreword

The microelectronics industry has used Pb-Sn solders for a variety of inter-connection needs, practically from its inception. It is not clear why Pb-Sn solders became the predominant solder material, although there is specula-tion that it was a carryover from the "electrical" industry, in which Pb-Sn solders were used to solder copper wires. Interestingly, another sector that used relatively large amounts of Pb-Sn solder was the plumbing industry. Following U.S. Federal regulations in 1988 requiring the removal of lead from potable water systems, the use of lead-bearing alloys to solder copper pipes and other fixtures was terminated. In the 15 years since then, our potable water systems have become cleaner than they were before.

The move toward using lead-free solders for interconnections in the microelectronics industry has been gaining momentum since at least the early 1990s, with the major Japanese companies leading the way. Today, several Japanese companies offer consumer electronics assembled with lead-free solders. Europe has also been pursuing the goal of making environmentally friendly electronics a reality, with the WEEE initiative. This is regulatory in nature, and will definitely bring about permanent changes.

Assemblers of electronic products are dispersed across the globe, with heavy concentrations in Asia, the Americas, and Europe, including Eastern Europe. Distribution of the relevant scientific and technological knowledge to this global industry is imperative for the smooth transition to lead-free interconnections not only in microelectronic devices but also in optoelectronic and MEMS devices.

The timing for this comprehensive book on lead-free solders could not be better, as it is needed and wanted. It offers several features that make it useful to the broad spectrum of readers, including professional engineers, research scientists, and graduate students. There is an optimal balance between the basic theories governing materials behavior and practical information necessary for the engineer in charge of the production line. The range of topics covered is indeed comprehensive, enabling the reader to use this book as a reference for understanding lead-free solder alloy behavior, alloy selection, alloy design, and alloy reliability. Each chapter has been written by an expert in that field—authors who not only are well versed in the applicable theories, but also have significant practical experience, both of which they share with the reader. The contributors are to be commended for the meticulous and thorough manner in which they have presented the information so that it can be of maximum value to the readers.

Lead-Free Soldering in Electronics: Science, Technology, and Environmental Impact should become an essential volume in the libraries of all individuals and companies that are involved in soldering. It should also be in the reference sections of all university libraries, so that it can be accessible to students. There is no doubt that this book will be very well received by the industrial and scholarly community, as they will be privy to very valuable knowledge—something that money alone can never buy.

Guna Selvaduray, Ph.D.
Chemical and Materials Engineering Department
San Jose State University
San Jose, California, U.S.A.

Preface

Metal interconnections have a long history, dating back to when mankind started to use metals 5000 years ago. The use of tin (Sn) soldering started in the Mesopotamian Age. It is an interesting fact that our ancestors used pure tin to solder a small silver grip to a copper bowl. This naturally produces a solder layer of tin with small amounts of silver and copper by dissolution from the substrates, which is the new standard of soldering today. Although our ancestors might not have used lead (Pb), because it was one of the precious metals in the Mesopotamian Age, we finally came back to the ancient alloy after 5000 years.

The solder—typically Sn-37wt%Pb eutectic alloy—developed with it has long provided many benefits, such as easy handling, low melting temperatures, good workability, ductility, and excellent wetting on copper and its alloys. Nowadays, soldering technology has grown to be the indispensable technology for the interconnection and packaging of virtually all electronic devices and circuits. Lead-bearing solders, especially the eutectic or near-eutectic Sn-Pb alloys, have been used extensively in the modern assembly of electronic circuits. Increasing environmental and health concerns about the toxicity of lead, however, as well as the possibility of legislation limiting the usage of lead-bearing solders, have stimulated substantial research and development efforts to find alternatives—that is, to find lead-free solder alloys for electronic applications.

Although several commercial and experimental tin-based lead-free solder alloys exist, none meets all standards, which include the required material properties (e.g., low melting temperature, wettability, and mechanical integrity), good manufacturability, and affordability. Current processing equipment and conditions (involving fluxes) have been optimized for Sn-Pb solder alloys over the last 30 years. The development of proper alloy compositions for the new solder systems—with suitable fluxes and assembly processes for lead-free solders—is also needed. In addition to the practical usage of lead-free solders, however, we need scientific information that enables us to understand the various phenomena occurring in electronics packaging employing lead-free solders. The development of a lead-free solder alloy that has all the aforementioned desirable properties and that allows easy assembly will be a formidable task, until we establish its scientific basis.

We currently have Sn-Ag-Cu alloy as the new standard. This alloy has many benefits even compared with Sn-Pb eutectic alloy: it has good manufacturability, good mechanical properties, and stable interfaces with most metallic substrates and surface finishes, as well as nontoxicity. Nevertheless, there are several key drawbacks of this alloy, including high melting temperature, slightly high hardness, and formation of solidification defects.

To address these issues, we need to make clear the scientific bases of lead-free solders as well as their surrounding technologies. This book summarizes the scientific and technological aspects of lead-free soldering in the following categories: the background and requirements in selection, properties of alloy candidates, metallurgical aspects, mechanical properties, plating technologies, processing technologies, and evaluation methods for solders and PWBs. We hope that *Lead-Free Soldering in Electronics: Science, Technology, and Environmental Impact* can serve as a valuable source of information to those interested in environmentally conscious electronics packaging.

I would like to take this opportunity to thank all the contributors—without their contributions, this book could not been accomplished. I wish to thank everyone who kindly supplied valuable information, photographs, and other materials on this new technology. I thank Mr. Taisuke Soda for his continual encouragement during the planning and preparation of this book and Ms. Karen Kwak for editing it with patience. I would also like to thank my colleagues and students in my laboratory, who supported me while I completed work on this book. Finally, I dedicate the book to my wife Miho, my daughters Kaoru and Midori, and my son Kohe, who allowed me to work many weekends on it.

Katsuaki Suganuma

Contents

Contributors

Shinichi Fujiuchi, M.Sc. NPI Center, Engineering No. 2, Sanmina-SCI Systems Japan, Ltd., Shiga-ken, Japan

Carol Handwerker, Ph.D. Metallurgy Division, National Institute of Standards and Technology, Gaithersburg, Maryland, U.S.A.

Masao Hirano, Dr.Eng.* Sensing Devices and Components Div. H.Q., Omron Corporation, Tokyo, Japan

Ursula Kattner Metallurgy Division, National Institute of Standards and Technology, Gaithersburg, Maryland, U.S.A.

Kilwon Moon Metallurgy Division, National Institute of Standards and Technology, Gaithersburg, Maryland, U.S.A.

Hidemi Nawafune, Ph.D. Chemistry of Functional Molecules, Faculty of Science and Engineering, Konan University, Kobe, Japan

---

* *Current affiliation*: Kyoto Prefecture Adviser (Technology Issues), Kyoto, Japan

Kay Nimmo, B.Sc. Research, Soldertec Global at Tin Technology Ltd, St. Albans, Hertfordshire, United Kingdom

Tetsuro Nishimura Marketing and Engineering, Nihon Superior Co., Ltd., Osaka, Japan

William J. Plumbridge, B.Sc., M.Sc., Ph.D., M.A., F.I.M. Department of Materials Engineering, The Open University, Milton Keynes, United Kingdom

Katsuaki Suganuma, Dr. Institute of Scientific and Industrial Research, Osaka University, Osaka, Japan

Hirokazu Tanaka, Ph.D. Reliability Engineering Department, Technical Development Headquarters, ESPEC Corp., Tochigi, Japan

Yun Zhang, Ph.D. Research and Development, PWB Materials and Chemistry, Cookson Electronics, Jersey City, New Jersey, U.S.A.

LEAD-FREE SOLDERING IN ELECTRONICS

1
Introduction to Lead-Free Soldering Technology

Katsuaki Suganuma
Osaka University, Osaka, Japan

1.1 INTRODUCTION

Environmentally friendly technologies found within the industries at the forefront of the age of electronics and automobiles are destined to become one of the key facets of technology for the new era. Many of the products that have been created within these fields have contributed to the affluence of human life, but at the same time have carried substances with a very high environmental load stemming from energy consumption and the discharge of noxious substances into the environment. To counter these problems, the home appliance recycling law has been enforced in Japan since April 2001. This is not only limited to Japan, but also extends to many of European and Asian countries in 2003. Thus worldwide recycling of home electronics has just begun. It will, however, take some time to determine the effects produced from recycling. Unfortunately, there have already been reports in the news of an increase in illegal dumping, and concern has surfaced regarding the lack of arrangements in the social systems surrounding recycling. Besides the promotion of recycling of home appliances, a massive amount of discarded electrical appliances is accumulating every year within the narrow confines of these countries. Toxic heavy substances contained in the electrical appliances, i.e., lead, mercury, hexavalent chromium, and cadmium, inhibit clean treatments of them in recycling and increase recycling cost. Although the removal of such toxic heavy substances from home electronics wastes is technologically possible, which is sometimes called "the end-pipe method," it creates a lot of trouble in our social systems. Considering this situation, rather than only waiting for the establishment of recycling, we must create our own models right in this century.

Sometimes it has been pointed out that lead in solder occupies only about 3% of the total lead consumption on average. Although the amount looks small, one must note that the other lead is in batteries, which have been recycled. Once solder is on a printed wiring board (PWB), it goes to landfills after all. Every year, we are accumulating lead in metallic form in our landfills. There is a simple calculation of the annual amount of lead in solder used for four typical home appliances going to landfills in Japan. The total number of these home electronics wastes is approximately 20,000,000 per year. Let's say one unit contains 10 g lead, and the total lead accumulated a year becomes about 200 t. One decade becomes 2000 t, two decades become 4000 t, ... Even worse, electronics wastes are not only the four appliances, but also a huge number of other products. The short-lived products such as cellular phones and personal computers (PCs) are our special concerns. There is no space for making new landfills in a small country. This is not only the situation for Japan, but also for many European and Asian countries, and even for many of the states in America. This is the reason why we are moving to lead-free technology in electronics packaging. If leading-edge industries are to maintain their position at the forefront of the new century, they must not be contented to merely offer individual products incorporating outstanding functions. The real key to their success will be their ability to reduce the environmental load one by one to fit to the current human life by the elimination of toxic substances and not by "the end-pipe method."

1.2 REGULATORY TRENDS FOR LEADED SOLDER

Legal regulations on leaded solder began early in 1990s with the U.S. "Lead Exposure Reduction Act" S2637 and S729. The proposals in these bills created a sudden burst of interest worldwide in developing lead-free solder. Movements to develop lead-free solder went forward in the United States under the NCMS Lead-Free Soldering Project [1] (NCMS: National Center for Manufacturing Sciences, 1994–1997) and in the EU under the IDEALS Project [2] (BRITE/EURAM 95/1994; Improved Design Life and Environmentally Aware Manufacturing of Electronics Assemblies by Lead-Free Soldering). In Japan, this movement was connected to the inauguration of the Lead-Free Soldering Research Council (1994 to 2000) within the Japan

Institute of Printed Circuits (currently, the JIEP: the Japan Institute of Electronic Packaging). While America has once moved away from legislating the changeover to lead-free solder due to technological difficulties, EU has included the adoption of lead-free solder in the movement to legislate disposal of all electronic products in its proposal for waste electrical and electronic equipment (WEEE). The final draft of the WEEE directive (in June 2000) deals separately with the recycling of hazardous elements, dividing recycling into two parts: WEEE and the RoHS (restriction of the use of certain hazardous substances in electrical and electric waste). This draft once proposed regulating lead in solder in 2008, but in May 2001, the EU parliament voted to move up the target date to 2006. The EU finalized this directive on February 13th in 2003, and the target date was fixed to July 1st in 2006 [3].

Table 1.1 summarizes the outline of the directives. Electronics equipments to be targeted are categorized into 10 groups. Almost all of the electronics products are involved in these categories except for space and military uses. Two categories, i.e., medical and monitoring products, are exempted for RoHS. The recycling and reuse rates are very high ranging from 60% to 80% in weight. If products contain those elements listed in the table, those products must be treated properly before disposal by the producer. All circuit boards larger than $10 \, cm^2$ in area and all circuit boards of mobile phones are involved.

In RoHS, lead, cadmium, mercury, hexavalent chromium, and two brominated frame-retardant materials are defined as prohibited hazardous substances. There are, however, several exemptions for each substance based on the possibility of alternatives. If there is no technological possibility for the alternatives, the substances can be used. Nevertheless, the producers must remove or make appropriate treatment before their disposal. Every 4 years, the EU reevaluates these exemptions by taking both technological developments and environmental impacts into account. As for a solder material, high-temperature solder that has lead content higher than 85 wt.% is exempted due to a technological viewpoint. Main infrastructures for telecommunication systems and certain kinds of servers are also exempted by the same reason. One needs to note that the end limit for servers is the year of 2010 and may be similar for other exempted products.

Within the EU, Denmark and Sweden are separately studying their own legislation. Denmark had already enacted a law at the end of the year 2000, but one item exempts "electrical parts." These "electrical parts" are thought to include all mountings. Sweden has been studying a long-term plan to eliminate the use of all lead by the year 2015–2020. However, the details of these regulations in each country will be restricted by the enactment of WEEE and RoHS finally.

In Japan, however, the subject of legislation to regulate lead solder has not yet been taken up. However, the control of lead is being strengthened through such means as review of the water quality standards concerning lead,

Table 1.1 Summary of WEEE/RoHS

Categories	10 categories: Large household appliances, small household appliances, IT/telecommunication equipment, consumer equipment, lighting equipments, electrical and electronic tools, toys/leisure/sports equipment, medical devices, monitoring/control instruments, automatic dispensers
Recovery/recycle by December 31, 2006	80%/75%: large household appliances and automatic dispensers 75%/65%: IT and consumer equipment –/80%: gas discharge lamps 70%/50%: others
Obligation to report	Member states shall send the report to the Commission
Treatment	The following substances and components have to be removed: Capacitors with PCB, components with Hg, batteries, PWBs of mobile phones and of other devices greater than 10 cm^2, toner cartridges, plastic containing brominated flame retardants, asbestos components, cathode ray tubes, CFC/HCFC/HFC/HC, gas discharge lamps, LCDs greater than 100 cm^2, external electric cables, components with refractory ceramic fibers, components with radioactivity, electrolyte capacitors with substances of concern (height >25 mm, diameter >25 mm)
RoHS; since July 1, 2006	Prohibition of use: Pb, Hg, Cd and Cr^{6+} PBB, PBDE: Exemption for Pb: High-temperature solder (Pb > 85%), servers/storage and storage array systems (until 2010), network infrastructure equipment for switching/signaling/transmission and network management for telecommunication, electronic ceramic parts, lead in glass of cathode ray tubes/electronic components/fluorescent tubes, alloying element in steel < 0.35%Pb/in Al, < 0.4%/in Cu, < 4%, spare parts for the repair/reuse put on market before July 1, 2006

RoHS: medical devices and monitoring/control instruments are not included.

strengthening amendments to disposal laws, and the enactment in April 2001 of the home appliance recycling law originating in 1998. Unless electronic devices containing lead are disposed of properly, they can no longer be discarded.

1.3 GLOBAL LEAD-FREE SOLDER DEVELOPMENT PROJECTS

In the United States, as soon as lead-free solder legislation was proposed, the lead-free solder project headed by the NCMS initiated research and development of lead-free solder in a program lasting 4 years. The results of the project have been made available in a database and offer information on such matters as modifying equipment and processes for selecting alternative materials [1].

The project initially selected for study 79 types of alloys considered at the time to be potential candidates for use in lead-free solder. Basic attributes considered included toxicity, resource availability, economic feasibility, and wetting characteristics. The selection process narrowed the field down to the final seven alloys, and these received secondary evaluation for reliability and ease of mounting manufacturing. Evaluation of the individual alloys did not result in the final selection of a single candidate, but three alloys, Sn–58Bi, Sn–3.5Ag–4.8Bi, and Sn–3.5Ag, were recommended as candidates. Screening comments indicated that the Sn–58Bi eutectic alloy was not suitable for use as standard solder due to the scarcity of Bi resources. However, since this material can be used for mounting at less than 200°C, and has chalked up a 20-year plus record of use in mainframes, this solder was deemed suitable for special applications.

These results were used to construct a database on lead-free solder that includes the information in these tables along with other items such as (1) recommended applications for lead-free solder, (2) alloy composition guidelines reflecting price and availability, (3) database of the 7 selected alloys and comparison with Sn–Pb eutectic alloy, (4) data on the characteristics of the other 70 eliminated alloys, (5) optimal process conditions using various test PWBs, (6) strength evaluation and metallurgical reaction analysis for the selected alloys and various surface mounting process reactions, (7) predicted life (using NCMS Project proprietary life prediction software) and thermal fatigue evaluation for 4 of the selected alloys, and (8) assessment of non-toxicity and alloy composition. Some of the data obtained in this project are shown in Chapters 2 and 5.

After the NCMS Project had finished, the move to legislate the use of lead-free solder was terminated, and research fervor cooled. However, Japanese progress in practical applications and EU legislation gave impetus to the creation of the National Electronics Manufacturing Initiative (NEMI) in May 1999, and the NEMI Task Force was organized to handle lead-free mounting. The major objectives of this group's activities are as follows.

(1) Obtain the capacity for manufacturing lead-free products in 2001 with a view to eliminating all lead in 2004.
(2) Clarify the parts, materials, and process conditions that can be used for lead-free solder mounting.
(3) Select a single main candidate from the Sn–Ag–Cu family of solder.
(4) Obtain cooperation of the various manufacturers of parts, materials, and equipment to confirm the smooth transition to high-temperature mounting to maximum temperatures around 260°C.
(5) Establish evaluation standards for lead-free processes.

This project carried out research on process and material conditions for lead-free mounting from the point of view of such items as reflow soldering, wave soldering, wash, and repair. Companies such as Celestica, Compaq, Delphi/Delco, HP, Motorola, Intel, IBM, NIST, Nortel Networks, Solectron, Visteon, and others participated in the activities. In addition, NEMI along with groups such as EIA (Electronic Industries Alliance), IPC, and JEDEC (Joint Electron Device Engineering Council) cooperated on duplicating the contents of those activities, performing investigative research and carrying out experimental development. The project was completed in January 2002, and some of the groups kept on research works, focusing both on whisker growth and test methods and on reliability tests.

In the EU, the IDEALS Project was carried out from May 1996 to April 1999 with the participation of companies such as Philips and Siemens [2]. The two main objectives of the project were to clarify the process window (range for process conditions such as thermal profiles) and to confirm reliability in practical use for alloys such as Sn–Ag–Cu, Sn–Ag–Bi, and Sn–Cu. Simultaneously, they also developed VOC-free flux. In the area of reparability, flux-containing Sn–Ag–Cu and Sn–3.5Ag had absolutely no problems, and the project was able to confirm the viability of both manual and automatic repairs.

The main conclusions of the IDEALS Project can be summarized as follows.

(1) The practical application of lead-free solder is technically and industrially feasible.
(2) Sn–Ag–Cu(–Sb) alloys can be used in a wide range of applications.
(3) The process window for wave soldering is roughly the same as for conventional solder.
(4) VOC-free flux has been developed.
(5) The process window for reflow soldering is compatible with most parts, but problems occur in one sector, from 225°C to 230°C.
(6) Sufficient reliability can be attained.

In addition to completing the project, Philips succeeded in bringing to mass production from the end of 1999 an Sn–Ag–Bi single-side wave soldered PWB for lighting utensils, and it had already delivered 1 million units by March 2000.

After IDEALS project, many research works have been performed in individual countries. From 2002, two major activities had again arisen under the support of EU. One is the large academic project called COST351, and the other is the industrial research and development carried out by the industrial consortium Interconnection Materials for Environmentally Compatible Assembly Technologies (IMECAT). The COST351 project aims to increase the basic knowledge on possible alloy systems that can be used as lead-free solder materials and to provide a scientific basis for a decision on which of these materials to use for different soldering purposes in order to replace the currently used lead-containing solders in the future. More than 40 universities and research organizations participate in this project. In contrast, the objective of the IMECAT project is to focus on the development of lead-free solders and adhesives (ICA: isotropically conductive adhesives, NCA: non-conductive adhesives) and the application of these materials for electrical interconnection and assembly for a wide variety of applications. Nine companies from five different EU member states are involved in this project.

In Japan, two years of research were completed on "Research and Development for the Standardization of Lead-Free Solder" supported by NEDO (New Energy and Industrial Technology Development Organization) as a Japanese national project in March 2000 [4]. Evaluation and development were carried out by two working groups, JEIDA [Japanese Electronic Industries Development Association (JEITA*)] and JWES (The Japan Welding Engineering Society) under the auspices of the NEDO organization the Japan Environmental Management Association for Industry. The main role undertaken by the JWES welding working group was in regard to evaluating material characteristics for lead-free solder candidate materials and investigating standard evaluation methods. The main role assumed by JEIDA was concerned with mounting characteristics and reliability evaluation with the aim of promoting practical application. Japanese Electronic Industries Development Association also cooperated with Electronic Industries Association of Japan (EIAJ, currently JEITA) in evaluating compatibility of various electronic parts and evaluating lead-free surface processes.

The following items sum up the JEIDA input.

Overall:

(1) Grasped the various characteristics of lead-free solder. Any type of solder has some limitations, but all are feasible for practical application.

Reflow:

(2) There are no significant problems with migration or insulation characteristics, but wettability is somewhat inferior to Sn–Pb eutectic solder.

(3) The greater the amount of bismuth in the alloy, the worse the compatibility with lead.

*Japan Electronics Information Technology Industries Association (JEITA) is a body inaugurated in November 2000 by combining Japan Electronic Industries Development Association (JEIDA) and Electronic Industries Association of Japan (EIAJ).

Wave:

(4) Liftoff generation is accelerated by the presence of bismuth quantities, but liftoff may occur even without bismuth, depending upon the conditions.

(5) Lead-free solder is superior to Sn–Pb eutectic solder in regard to creep and thermal cycles.

After the NEDO project, a number of new projects were initiated during 2000 and 2001. One of those was the IMS project "Developing environmentally friendly next-generation joining technology" headed by Hitachi, Ltd. The project was budgeted by the Ministry of Economy, Trade, and Industry, and in the year 2000, participants included Hitachi, Sony, Sharp, Oki Electric Industry, and some universities and national research laboratories. In 2003, the organization was expanded even further internationally involving Korea and some of the European countries. This project targets technological evaluation of lead-free solder regarding reliability and mounting characteristics, clarifying the biological impact of the component elements, and also establishing high-level mounting technology that will be safe and environmentally friendly into the future.

The NEDO projects on "Standardization of necessary test methods for solder connections that reduce the environmental load" were carried out in the fiscal year of 2001 by the JWES welding group. As can be seen from the title, the project targets standardization of test methods for solder materials. Meanwhile, JEITA has initiated a project to standardize reliability and mounting evaluation from the parts aspect in the same year for a 3-year project until 2004. They are also intensively working on whisker growth and evaluation methods, reliability test methods, and migration evaluation methods.

The Japan Institute of Electronics Packaging (JIEP) initiated the "Low-temperature lead-free solder development project" as a 2-year plan to intensively investigate solder capable of low-temperature soldering from 2001 to 2002 [5]. Items include solder materials such as Sn–Zn and Sn–Bi, and future aims extend to presenting available specifications to all persons involved in lead-free soldering as well as constructing various databases. The "low-temperature" target is striving for soldering temperatures below those of current Sn–Pb eutectic solder, i.e., 220°C. After the JIEP low-temperature soldering project, this has been followed by the JEITA to expand the usage of Sn–Zn soldering technology in industries since 2003.

1.4 INDUSTRIAL TRENDS

The major Japanese electronics manufacturers lead the world in mass production that incorporates lead-free solder. The world's first mass-produced product incorporating lead-free solder was the Panasonic compact MD player in the autumn of 1998 (Fig. 1.1) by using Sn–Ag–Bi–In solder batch reflow. Basically, they are using Sn–3.5Ag–0.5Cu for many products of reflow and wave soldering. As for low-processing temperature requirement, they are using Sn–3.5Ag–(3–8)In–0.5Bi alloys. Figure 1 also shows a note PC and its main PWB soldered with Sn–3Ag–8In–0.5Cu. Figure 1.2 shows the world's first PWB of a videotape recorder from Panasonic mass-produced by wave soldering with Sn–0.7Cu with a little amount of nickel. In the middle of 2002, they have almost more than 200 new products into lead-free soldering, and in March 2003, they have almost finished converting to lead-free assembly.

NEC Corp. began using Sn–3.5Ag–0.7Cu solder to manufacture pagers (pocket bells) from the end of 1998. Figure 1.3 is the note PC and its PWB soldered with Sn–8Zn–3Bi first released worldwide by NEC in 1999.

Hitachi has been actively promoting the changeover to lead-free solder since 1999 and has announced that approximately half of the products it manufactures in Japan were using lead-free solder by mid-2000. The company is performing reflow for many products with Sn–Ag–Cu solder and is not only using lead-free solder, but has also changed over to halogen-free PWBs. They have also almost finished converting to lead-free assembly in 2003.

Sony completed its adoption of lead-free solder in VTRs and camcorders in March 2000 and has now begun the changeover in overseas of many products such as TV manufacturing. The solder used is Sn–3Ag–0.5Cu.

Toshiba, Fujitsu, Sharp, Epson, Victor, Pioneer, Richo, and many other major electronics companies in Japan also started manufacturing a

Figure 1.1 Compact MD and note PC (courtesy of Panasonic). Solders used are Sn–2.7Ag–3In–0.5Bi and Sn–3Ag–8In–0.5Bi, respectively.

Figure 1.2 World's first wave soldered videotape recorder and its one-sided board soldered with Sn–Cu(–Ni) (courtesy of Panasonic).

large number of lead-free products by the year of 2000–2001. It is interesting to note that, in 2001, Japanese, European, and American companies released lead-free soldered cellular phones. They are all soldered with Sn–Ag–Cu alloys of different compositions.

During the year of 2000, the use of lead-free solder mountings on automobiles was begun. Because automotive design places its highest priority on safety, an exception was made for automobiles in the EU's ELV (end-of-life) directive enacted in 2000. However, there are an enormous number of automobiles discarded annually, similar to the situation of home appliances, and so harmful elements such as lead need to be replaced as soon as possible, and each company has plans for eliminating such substances. Nissan has led the field in combating this situation by introducing a Sn–Ag–Cu lead-free

Figure 1.3 World's first note PC and its circuit board soldered with Sn–8Zn–3Bi (courtesy of NEC).

solder mounting for its mass-produced keyless entry system PWB, adopted from the summer of 2000. This is probably a global first public proclamation of adopting lead-free solder for automotive use. The PWB for that system is shown in Fig. 1.4. In America, too, Ford automobiles have lead-free solder mountings on safety lock systems installed from the end of 2000.

In the ways noted above, Japan leads the world in adopting lead-free solder for mass production. There was some apprehension that companies would not continue moving forward on the issue, but all trends seen since have indicated no slowing of the transition to lead-free solder. Rather, there seems to be even greater haste to make progress. The awareness of the Japanese companies is also shared by American and European companies, and information from the major electronics manufacturers indicates that all are going full speed ahead in promoting lead-free solder mounting.

1.5 TYPES OF LEAD-FREE SOLDERS IN USE

In the past, Sn–Pb eutectic solder is not the only solder that has been used; a variety of solder compositions have been employed, and even with lead-free

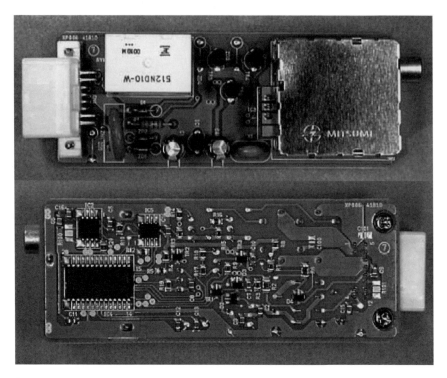

Figure 1.4 Keyless entry system of automobile wave soldered with Sn–Ag–Cu (courtesy of Nissan).

solder, we must be sure to use the proper material in the proper place. In particular, Sn–Pb solder has been used to cover a wide temperature range, but the need has arisen to substitute various other alloy compositions. Table 1.2 presents a summary of the composition and characteristics of current lead-free solders that can be cited, along with notes about their usage. Since the details of scientific and practical aspects of those alloys are mentioned in the following chapters, the current usages are briefly summarized in this chapter.

Among the various candidates, the Sn–Ag–Cu family of solder is the strongest candidate to become the standard lead-free solder. Research in Japan, Europe, and the United States indicates that this alloy is extremely stable and, accordingly, is considered able to meet globally acknowledged standards. However, the compositions being recommended in Japan, Europe, and the United States have slight differences. Information obtained through simulations and high-precision scientific trials seem to call for a eutectic composition of around Sn–3.5Ag–0.7Cu. One obstacle for the Sn–Ag–Cu alloy is

Table 1.2 Lead-Free Solder Candidate Alloys in Use and Their Cautions

Processes	Alloys	Notes
Wave soldering	*Sn–Ag family* Sn–3.5Ag Sn–(3–3.9)Ag–(0.5–0.8)Cu *Sn–Cu family* Sn–0.7Cu With some additions: Ag, Au, Ni, Ge, In, etc.	Compatibility with Sn–Pb plated components. Fillet-lifting and land-lifting. For one-sided PWBs, Bi can be added, although the compatibility with 42 alloy should be noted.
Reflow soldering		
High temperature	*Sn–Ag family* Sn–3.5Ag Sn–(3–3.9)Ag–(0.5–0.8)Cu Sn–(2–4)Ag–(1–6)Bi With 1–3%In	Heat resistance of components and PWBs. One needs to control temperature distribution on PWBs. Compatability of Bi with Sn-Pb plating.
Intermediate temperature	*Sn–Zn family* Sn–9Zn, Sn–7Zn–Al Sn–8Zn–3Bi *Sn–Ag–In family* Sn–3.5Ag–(6–8)In–0.5Bi	Corrosion should be noted, especially for chlorine and for humidity. Barrier plating such as Au/Ni can improve high-temperature stability. Compatibility with certain kind of plating.
Low temperature	*Sn–Bi family* Sn–57Bi–(0.5–1)Ag	Compatibility with Sn–Pb plated components.
Hand and robot soldering	*Sn–Ag family* Sn–3Ag–0.5Cu Sn–Cu, Sn–Bi families	Compatibility among different solders and fluxes.

that raising the melting creates a corresponding rise in soldering temperature. Sn–Pb eutectic solder has a melting point of 183°C, and a reflow temperature of 230°C is quite sufficient for manufacturing. However, the Sn–Ag–Cu eutectic alloy has a melting point of 216°C, which is 33°C higher than conventional solder. As a result, a reflow temperature of 230°C leaves too narrow a margin and results in almost complete lack of success, and so raising the reflow temperature becomes unavoidable. Higher reflow temperatures demand better thermal resistance for semiconductors as well as other parts and PWBs. Concern for this very problem led the NEMI Project to require 260°C thermal resistance for the industry. In addition to necessitating higher thermal resistance of parts, this approach also necessitates higher precision in mounting temperature control. Urgent development of mounting parts is underway in an effort to limit thermal distribution on PWBs to a minimum and thus make it possible to handle items such as large PWBs as well as permit the coexistence of large and small parts. Devices with small reflow thermal distribution are already being marketed, and optimized design and parts layout should make reflow temperatures of 240°C feasible even for large PWBs. Recent improvement of reflow ovens enables the stable production of many products with Sn–Ag–Cu. Fig. 1.5 shows a cellular phone soldered with Sn–3Ag–0.5Cu, which is one of the JEIDA recommended alloy.

The addition of Bi to Sn–Ag solder lowers the melting point and improves wettability, making this a very attractive option. On the other hand, a severe defect called liftoff forms when using the Bi additive with wave soldering for double-side PWBs, urgently necessitating suppression countermeasures as well as the ability to ascertain conditions causing the defect. It is known that adding more than a certain percentage of Bi results in a weak-

Figure 1.5 Cellular phone soldered with Sn–3Ag–0.5Cu (courtesy of Fujitsu).

ened interface with 42 alloy (an alloy of Ni and Fe used for items such as IC lead frames). This interface already has a low initial strength, which is compounded by striking degradation resulting from thermal fatigue. The cause of this low interface strength has not yet been worked out and awaits further research. Adding Bi to solder is known to cause compatibility problems with Sn–Pb plated parts. Compatibility difficulties stem from a low melting point phase forming in the interface with Sn–Bi–Pb, and this phase then generates a melting reaction at less than 100°C. Therefore care must be taken when using Bi additives in solder used with plating containing lead.

Sn–Bi alloys have a eutectic point of 139°C and are attractive as solder because they can be used for mounting at temperatures below 200°C. Past lack of ductility has been an obstacle, but it has been possible to improve ductility through the use of such additives as silver. Use of this alloy has made it possible, for example, to adopt Sn–Ag solder as a high-temperature solder making feasible hierarchical junctions (high-, mid-, and low-temperature multistage connections; step soldering) in mounting. Rather than requiring thermal resistance to exceed 100°C for most home appliance products, a better option would probably be to have this low melting point solder available for a wide range of products. Compatibility problems with Sn–Pb plating are an important concern, just as with Sn–Bi eutectic solder.

The Sn–Zn–Bi family of solder has a melting point of 189°C, making it an alloy with characteristics for mounting that are extremely close to the current Sn–Pb solder. However, the marked oxidation of Zn yields poor mounting characteristics, and so early on the United States and Europe gave up on adopting this type of solder. Japanese manufacturers, however, improved the flux, making it possible to use this solder for mounting with no inferiority at all to conventional technology in the atmosphere. In the future, the possibility exists for developing solder with higher reliability through such means as evaluating resistance to corrosion. Problems of compatibility with Sn–Pb plating are the same as with other solders containing bismuth. Currently, this solder contains a 3% bismuth additive, but the optimum composition needs to be reconsidered due to severe compatibility problems with Sn–Pb plating and the occurrence of phenomena such as liftoff. The amount of bismuth should probably be reduced to less than 3%. One example is Sn–9Zn eutectic alloy with a small amount of additives. Sn 7Zn alloyed with a few 10 ppm aluminum has been developed [6]. Since aluminum acts as a surface protection element, this alloying improves wetting properties without bismuth. One of the typical products from Fujitsu with Sn–Zn–Al solder is shown in Fig. 1.6. One of the serious concerns on Sn–Zn alloy solder is its poor corrosion resistance. The protective coating mixed in flux.

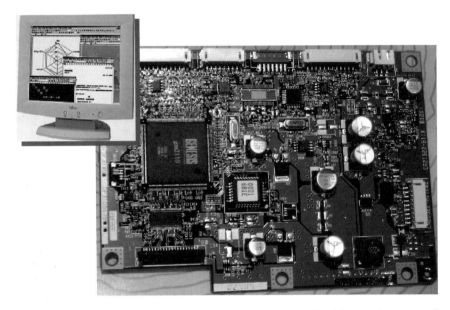

Figure 1.6 LCD and its main board soldered with Sn–7Zn–Al paste (courtesy of Fujitsu).

The reasonable price of the Sn–Cu family of solder makes it an essential alloy for wave soldering applications. However, there is some apprehension about the low reliability of Sn–Cu due to such problems as thermal fatigue, making this a less desirable choice for mountings that demand high reliability. Inferior wettability also makes it difficult to properly wet through holes, so this solder will probably need to be limited to such uses as wave soldering on single-side PWBs. Ongoing trials for reliability improvement techniques consist of subtle changes in composition such as adding a third element such as silver, gold, or nickel. For wave soldering of one-sided PWBs, Sn–Cu becomes one of the good choices. The higher melting temperature of this alloy at 227°C will not cause major changes in wave soldering temperature conditions compared to conventional Sn–Pb eutectic solder. However, due to concerns about PWB thermal resistance as well as compatibility problems with lead-plated parts, further improvements are needed.

Thus, up to now, we have a series of lead-free solder alloys with the accumulated database and know-how to handle them. Appropriate applications of these solders based on them can improve reliability of products resulting in longer lives for lead-free soldered products. These are the desirable products of the new era for which we are aiming.

REFERENCES

1. Lead-Free Solder Project Final Report. NCMS Report 0401RE96, National Center for Manufacturing Sciences, Michigan, 1997.
2. MR Harrison, JH Vincent. IDEALS: Improved design life and environmentally aware manufacturing of electronics assemblies by lead-free soldering. Proceedings of 12th Microelectronics & Packaging Conference. Cambridge: IMAPS Europe, 1999, pp 98–109.
3. Directive of the European Parliament and of the Council on waste electrical and electronic equipment (WEEE) and on the restriction of the use of certain hazardous substances in electrical and electronic equipment. Off J Eur Union, 13.2.2003.
4. Results of Investigative Research on Lead-Free Solder (00-base-17). Tokyo: Japan Electronic Industry Development Association, 2000.
5. Low Temperature Soldering. Low-temperature solder mounting technology development project of the Japan Institute of Electronics Packaging, Tokyo, 2002.
6. M Kitajima, T Shono, T Ogino, T Kobayashi, K Yamazaki, M Noguchi. Development of the tin–zinc–aluminum solder alloys. 2003ICEP Proceedings. Tokyo: IMAPS Japan/JIEP/IEEE CPMT Japan, 2003, pp 339–344.

2
Materials Science Concepts in Lead-Free Soldering

Carol Handwerker, Ursula Kattner, and Kilwon Moon
National Institute of Standards and Technology, Gaithersburg, Maryland, U.S.A.

2.1 INTRODUCTION

The metallurgical "life cycle" of a solder joint, whether lead-free or containing Pb, has multiple stages. First, for surface mount technology, the solder alloy is atomized into powder and fabricated into a paste. For through-hole technology, the solder alloy is melted in the wave soldering pot. Second, the temperature of the solder must be sufficiently low during reflow or wave soldering to avoid damage to the board and components but high enough to melt and wet the components and the board in a reasonable processing time. Third, the solder joint must solidify without forming defects that affect the joint's integrity, during and immediately after solidification, and during use. Next, the solder joint must be able to withstand the stresses imposed by use, including thermomechanical fatigue, thermal shock, vibration, and impact. Last, when the product is removed from service, the circuit boards with their solder joints become waste and are burned, buried, or recycled.

Over the last 10 years, research and development of lead-free solder alloys have focused on identifying alloys that meet specific criteria for manufacturing, reliability, toxicity, cost, and availability. These criteria can be understood by examining the metallurgy of and constraints on the "life cycle" of a typical solder joint with respect to manufacturing and reliability. (Issues of toxicity, availability, and cost are examined in other chapters.) In this chapter, we examine the metallurgy of solder alloys. This includes phase transformations in solders (including melting behavior, solidification pathways and interface reactions with substrate and component lead materials), wetting behavior, and mechanical properties (including thermomechanical fatigue). The metallurgical issues are illustrated using examples and data from a wide range of sources, including the NCMS Lead-Free Solder Project (US) [1–4], the IDEALS Lead-Free Solder Project (UK) [5–8], the NEMI Lead-Free Assembly Project (US) [9,10], various Japanese consortia [11–13], the National Institute of Standards and Technology (NIST) [14–17], and the open literature.

Laboratory tests for identifying phase transformations, wetting behavior, and mechanical properties can be quite successful in reducing the number of lead-free alloys to those most likely to be acceptable as replacements for eutectic Sn–Pb in circuit board assembly. Beyond an initial down-selection process based on "pass-fail" type criteria, no suite of laboratory experiments has been identified which can provide an accurate ranking of possible lead-free alloys. The challenges in using laboratory test results to identify the "perfect" solder alloy to replace Sn–Pb eutectic are illustrated by examining the formal, quantitative ranking process used by the NCMS Lead-Free Solder Project. The problem lies with finding laboratory tests or even experiments using circuit board test vehicles that accurately predict the thermomechanical fatigue life of lead-free solders for the full range of currently used components, circuit boards, and product conditions. In the chapters that follow, the current understanding of thermomechanical fatigue of lead-free solders is discussed, and the constraints imposed by the practicalities of manufacturing practice and product applications are addressed in greater detail.

2.2 PHASE TRANSFORMATIONS IN SOLDER ALLOYS

Phase transformations in solder alloys include melting, dissolution of and reaction with the board and component lead materials while the solder is in the molten state, solidification which includes issues of nucleation, precipitation on preexisting phases and metastable phase formation, interdiffusion, coarsening, and reaction with the substrate in the solid state, and changes to the solubilities and the distribution of phases as a result of thermomechanical fatigue. It also may include "tin pest," the transformation of beta to alpha tin at low temperatures, leading to a volume expansion of 23% and a catastrophic disintegration of solder joints [18]. In terms of a solder joint's "life cycle," some of these are clearly important and straightforward to analyze, such as melting and solidification behavior [14–17,19]. In terms of evaluating a lead-free solder alloy for commercial use, the other transformations are important only in how they affect the ability to fabricate and put into use reliable solder joints [20]. The relationships between these factors influencing reliability are therefore dependent on board and component materials, including surface finishes, thermal history in processing, and thermomechanical history in use, and will be discussed briefly in the Reliability Section below.

2.2.1 Melting Behavior

In choosing a lead-free solder as a replacement for Sn–Pb eutectic, the first place to start is a consideration of melting behavior. Because the behavior of lead-free solder alloys is being judged against Sn/Pb eutectic, it is useful to begin the discussion with an examination of the Sn–Pb phase diagram (Fig. 2.1) and the melting behavior of Sn–Pb alloys. The Sn–Pb phase diagram is characterized by two solid phases each with substantial solid solubility and a liquid phase. Further, the system is characterized by a simple eutectic with a significant depression of the liquidus temperature (T_l) by almost 50°C, from pure Sn at 232°C to the binary eutectic (Sn–37Pb) at 183°C. The microstructure on solidification is a mixture of Sn and Pb solid solutions.

The Sn–Bi, Sn–Ag, and Sn–Sb systems are typical of the types of melting behavior for lead-free alloys [5]. In the Sn–Bi diagram (Fig. 2.2), there is

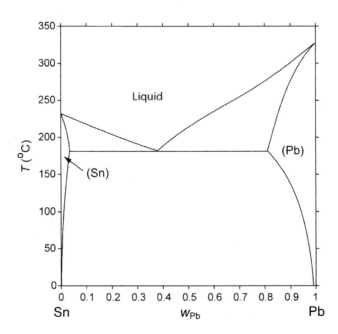

Figure 2.1 Sn–Pb phase diagram.

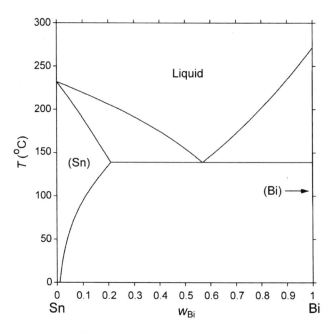

Figure 2.2   Sn–Bi phase diagram.

significant solid solubility of Bi in Sn, up to 22% Bi in Sn at the eutectic
temperature, 139°C. (All compositions are given in mass fraction × 100.) The
liquidus temperature decreases with increasing Bi concentration, from 232°C
at pure Sn to 139°C at 58% Bi. The solidus temperature decreases with
increasing Bi concentration, from 232°C at pure Sn to 139°C at 22% Bi. In
the Sn–Ag diagram (Fig. 2.3), there is negligible solid solubility of Ag in Sn.
The liquidus temperature decreases from 232°C to 221°C at 3.5% Ag. The
Sn–Sb system (Fig. 2.4) contains a peritectic at the Sn-rich side of the phase
diagrams, leading to an increase in liquidus temperature with increasing Sb
concentration.

There has been widespread desire on the part of the microelectronics
industry both 1) to keep the liquidus temperature as close as possible to
183°C, the eutectic temperature of Sn–Pb, in order to avoid changing manufacturing processes, materials, and infrastructure; and 2) to keep the solidus
temperature as close as possible to the liquidus temperature, to avoid fillet
lifting and ensure that the solidus temperature is significantly higher than the
solder joint's maximum operating temperature. Eutectics obviously meet the
second criterion. In terms of the temperature criteria, however, eutectic and
peritectic Sn-based alloys tend to fall into two temperature regimes. The high
temperature, Sn-rich eutectics are Sn–0.9Cu (227°C), Sn–3.5 Ag (221°C), Sn–

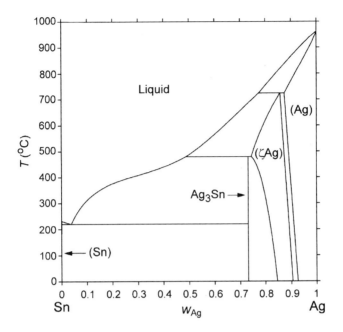

Figure 2.3   Sn–Ag phase diagram.

9Zn (199°C), and Sn–3.5Ag–0.9Cu (217°C). The low temperature eutectic and peritectic solders are Sn–58Bi (139°C), Sn–59Bi–1.2Ag (138°C), and Sn–In (120°C). The NCMS Lead-Free Project member companies selected solders with liquidus temperatures less than 225°C and with a pasty range (the difference between liquidus and solidus temperatures) to less than 30°C. The IDEALS and NEMI projects limited its candidate solders to eutectic and near eutectic, Sn-rich solders.

For Sn-rich solders, a simple linear equation can be used to estimate the change in liquidus temperature with composition for additions of Ag, Bi, Cu, Ga, In, Pb, Sb, and Zn to Sn [1]. For Ag, Bi, Cu, and Pb, the coefficients were derived from the slopes of the Sn–X (X = Ag, Bi, Cu, Pb) binary phase diagram line depicting liquidus temperature as a function of compositions for the Sn-rich liquidus, for In and Sb from the binary Sn–Y (Y = In, Sb) peri tectics, and for Zn and Ga from the extrapolations of the Sn-rich liquidus for dilute Sn–Zn and Sn–Ga alloys. The change in the liquidus temperature as a function of composition can be estimated as:

$$T_l = 232°C - 3.1*W_{Ag} - 1.6*W_{Bi} - 7.9*W_{Cu} - 3.5*W_{Ga}$$
$$-1.9*W_{In} - 1.3*W_{Pb} + 2.7*W_{Sb} - 5.5*W_{Zn} \tag{2.1}$$

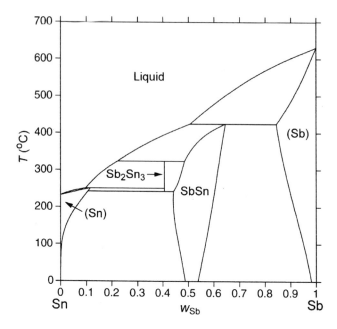

Figure 2.4   Sn–Sb phase diagram.

where the coefficients are in units of $°C$, and W_X is the amount of element X in mass fraction*100. Table 2.1 lists the composition limits over which this equation applies.

Accordingly, the maximum decrease from the melting point of pure Sn with additions of Ag and Cu is 15–16°C, in agreement with the measured ternary eutectic temperature in the Sn–Ag–Cu system of 217°C, as seen in Fig. 2.5 [14]. From Eq. (1), many alloy compositions with Bi, In, and Zn additions can be identified with liquidus temperatures of 183°C, the eutectic temperature of Sn–Pb eutectic solder. The problem with most of these alloys is that their solidus temperatures are significantly lower than 183°C. This issue of limiting the pasty range is particularly serious for through-hole joints: alloys with a large pasty range can exhibit separation between the solder fillet and the land on the board side during solidification. This phenomenon is known as "fillet lifting" in plated-through-hole joints.

Three compositions in the Sn–Ag–Cu system have been chosen as replacements for Sn–Pb eutectic solders. The preferred solder compositions are Sn–3.0Ag–0.5Cu (Japan), Sn–3.5Ag–0.9Cu (EU), and Sn–3.9Ag–0.6Cu (US). Figures 2.6a and b show the equilibrium fractions of different phases as a function of temperature for Sn–3.0Ag–0.5Cu and Sn–3.9Ag–0.6Cu (US). A

Table 2.1 Composition
Limits for Calculation of
Liquidus Temperature as a
Function of Composition

Tin-based compositions
(liquidus calculation,
mass fraction × 100)

Element	Weight %
Ag	< 3.5
Bi	< 43
Cu	< 0.7
Ga	< 20
In	< 25
Pb	< 38
Sb	< 6.7
Zn	< 6

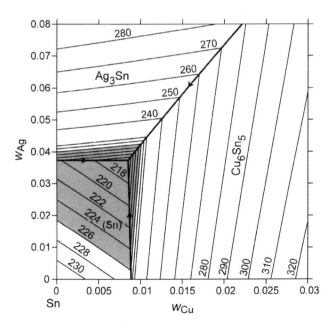

Figure 2.5 Sn–Ag–Cu phase diagram.

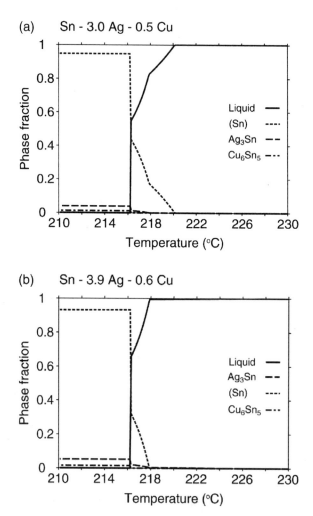

Figure 2.6 (a) Phase fractions for Sn–3.0Ag–0.5Cu as a function of temperature. (b) Phase fractions for Sn–3.9Ag–0.6Cu as a function of temperature.

comparison of the calculated fraction solid as a function of temperature for these three alloys, plotted with two other compositions (Fig. 2.7), illustrates an important point regarding the sensitivity of the melting behavior to changes in composition: The total fraction of intermetallic phases over wide composition ranges is small and difficult to detect using standard DTA systems. The "effective" liquidus temperatures measured will therefore be 217°C for a wide range of compositions. In reflow soldering, it is likely that this small fraction of intermetallic phase will have a correspondingly small

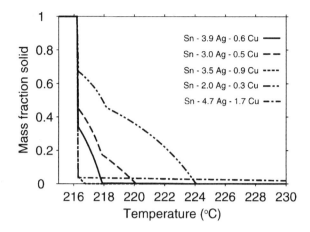

Figure 2.7 Phase fraction solid as a function of temperature for five different Sn–Ag–Cu alloys.

effect on solder flow and wetting, even if the solder in the joint never becomes completely liquid.

2.2.2 Solidification Behavior

Nonequilibrium Effects: The pasty ranges based on equilibrium phase diagrams are the minimum pasty ranges that will occur during solidification. Nonequilibrium segregation and metastable phase formation may extend these ranges. In systems that exhibit substantial changes in the solubility of elements in solid Sn during cooling, the amount of liquid present during cooling can be greater than predicted from the equilibrium phase diagram. Tin-based solder systems that exhibit this effect include Sn–Bi, Sn–In, and Sn–Pb. If diffusion in the solid does not establish the equilibrium solid composition at each temperature as a Sn-rich Sn–Bi solder cools, the remaining liquid becomes increasingly Bi-rich and will solidify at the eutectic temperature. If there is solid-state diffusion that maintains the equilibrium solid composition, the final liquid transforms to solid at the equilibrium temperature and composition. For a Sn–Bi solder with 6% Bi, the liquidus temperature is approximately 224°C and the equilibrium pasty range is approximately 26°C; with no diffusion in the solid, the pasty range can be as large as 85°C. In the NCMS Lead-Free Project, DTA measurements of Sn–6Bi detected a measurable fraction of eutectic liquid that solidified at 139°C.

This effect is illustrated in Fig. 2.8 for the ternary Sn–Ag–Bi system with calculations of the solid fraction as function of temperature and composition based on both the phase diagram "lever rule" and nonequilibrium solidifi-

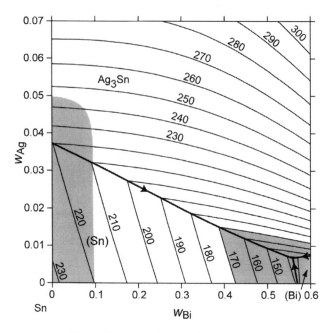

Figure 2.8   Sn–Ag–Bi phase diagram.

cation [1–4,15]. The liquidus projection of the ternary phase diagram is shown in Fig. 2.8, where the lines correspond to compositions with the same liquidus temperatures. Considering the composition Sn–3.5Ag–7.5Bi, the last liquid solidifies at 185°C based on the equilibrium phase diagram; however, as a result of segregation during solidification, some liquid is predicted to still be present until the ternary eutectic temperature is attained at 138°C. The amount of nonequilibrium liquid present depends on the cooling conditions and will be between the limits defined by the two curves for Sn–3.5Ag–7.5Bi in Fig. 2.9a and b.

Metastable Phase Formation: Another characteristic of many lead-free solder systems is the formation of nonequilibrium phases, as illustrated using the Sn–Ag–Cu system [14]. The Sn–Ag–Cu phase diagram, the calculated solidification path, and DTA results for the Sn–4.7 Ag–1.7 Cu are shown in Figs. 2.5, 2.10a and b, respectively. During cooling from the liquid state, the first phases to form are Ag_3Sn and Cu_6Sn_5 at 244°C as seen in Fig. 2.10a and b. At equilibrium, the remaining liquid should transform to a mixture of Sn, Ag_3Sn, and Cu_6Sn_5 at the ternary eutectic of 217.5°C. From DTA experiments (Fig. 2.10b), because solid Sn is difficult to nucleate, the liquid super-

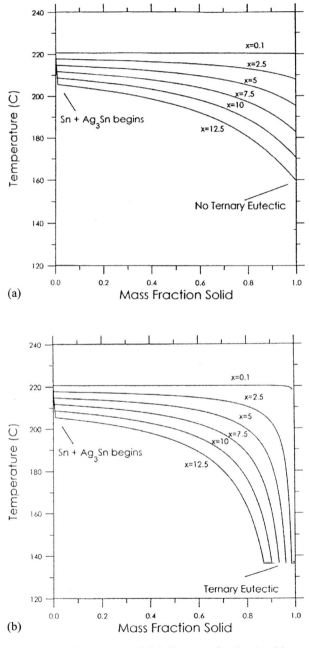

Figure 2.9 Lever and Scheil curves for Sn–Ag–Bi.

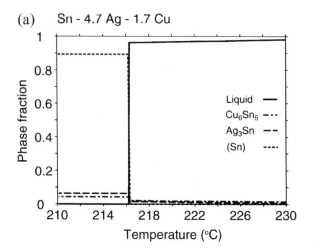

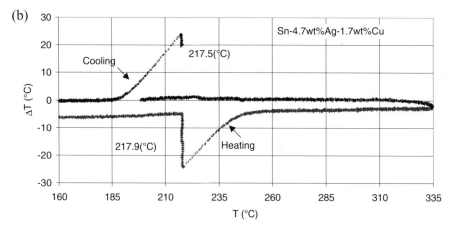

Figure 2.10   (a) Phase fractions for Sn–4.7Ag–1.7Cu as a function of temperature. (b) DTA heating and cooling curves for Sn–4.7 Ag–1.7 Cu.

cools while Ag_3Sn and Cu_6Sn_5 continue to form until the remaining liquid solidifies at 198.5°C. The heat of formation is released, leading to the solder self-heating to 217°C. This phenomenon is known as recalescence and is also exhibited in the Sn–Pb system. For the Sn–Ag, Sn–Cu, and Sn–Ag–Cu systems in particular, the existence of liquid below the eutectic temperature in the Sn–Ag, Sn–Cu, and Sn–Ag–Cu means that intermetallic phases form and coarsen in the liquid for significantly longer than expected from equilibrium behavior and the liquid becomes Sn-rich by the continued formation of the intermetallics. When the Sn phase in this Sn-rich, off eutectic liquid phase finally nucleates, the Sn phase grows dendritically. This is the origin of the observed microstructures, which contain tin dendrites, rather than a classic "eutectic" microstructure. A comprehensive experimental and theoretical study by Moon et al. on the Sn–Ag–Cu system provides more detailed discussion of the microstructures, melting behavior, and solidification behavior, which applies to Sn–Ag, Sn–Cu, and alloys of Sn–Ag–Cu with other alloy additions [14].

Fillet Lifting: A failure phenomenon for through-hole joints that occurs for some lead-free solders during solidification that does not occur for eutectic Sn–Pb is "fillet lifting." Fillet lifting, as shown in the micrograph in Fig. 2.11, is characterized by the complete or partial separation of a solder joint fillet from the intermetallic compound on the land to the shoulder of the through hole. This phenomenon was first identified in 1993 by Vincent and coworkers in the DTI-sponsored Lead-Free Solder Project, which attributed fillet lifting to the presence of the Sn–Bi–Pb ternary eutectic (98°C) due to Pb contamination of Bi-containing solders from the Sn–Pb HASL board finish [21–23]. This effect is now known to occur without Pb contamination for some lead-free solder alloys.

From the work by Suganuma, Boettinger et al., and others, fillet lifting has been found to be a result of "hot tearing," a mechanism that leads to relief

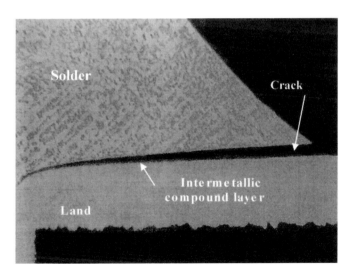

Figure 2.11 Optical microscope cross section of a through hole solder joint with Sn–3.5Ag–5Bi alloy.

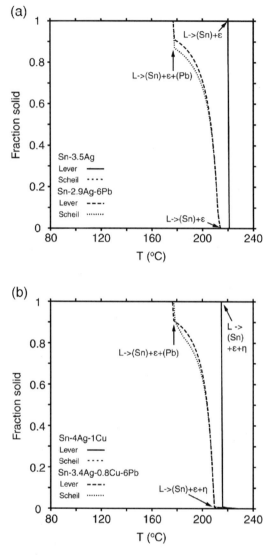

Figure 2.12 Lever and Scheil calculations for fraction solid as a function of temperature for Pb-free solders without and with Pb contamination at the six mass fraction level for (a) Sn–Ag and (b) Sn–Ag–Cu.

of thermally induced stresses *when the solder is between 90% and 100% solid* [24–26]. The differential shrinkage due to CTE mismatch between the board and the solder generates the stresses; at lower solid fractions, fluid flow occurs relieving the stresses. As the volume fraction of liquid decreases, the stresses are carried by the dendritic matrix and failures occur at the weakest point, the location with the highest remaining liquid fraction: the board-side intermetallic compound/solder interface. The tendency for hot tearing increases as the pasty range increases and the temperature difference between 90% and 100% solid increases and is typically worse for alloys with a large nonequilibrium pasty range, such as Sn–Bi or Sn–Ag–Bi.

In the NCMS Lead-Free Solder Project, the "hot tearing" hypothesis was tested by taking Sn–3.5Ag, an alloy that showed minimal fillet lifting, and transforming it into an alloy showing close to 100% cracked joints with the addition of 2.5 wt.% Pb. The addition of 2.5% Pb increased the pasty range from 0°C to 34°C. These results suggested that Pb contamination from Sn–Pb surface finishes would lead to fillet lifting in alloys that in their uncontaminated state showed little or no fillet lifting. Subsequent wave soldering experiments by Multicore, Nortel, and others exhibited fillet lifting in through-hole joints with Sn–Ag, Sn–Cu, or Sn–Ag–Cu solders and Sn–Pb surface finished components and/or boards. Thermodynamic calculations by Kattner (Fig. 2.12a and b) illustrate the effect of Pb-contamination on the pasty ranges of Sn–3.5Ag and Sn–Ag–Cu.

2.3 WETTING AND SOLDERABILITY

2.3.1 Wetting Balance Test Methods

Wetting of a liquid on a solid is determined by the relative energies of the liquid–vapor surface tension, the solid–liquid interfacial energy, and the solid–vapor interfacial energy. The thermodynamics of an alloy plays a central role in determining its intrinsic surface tension. Ohnuma and his colleagues have used thermodynamic parameters to predict the surface tension, the surface composition, and the viscosity of the Sn-based liquid solder as a function of composition, as shown in Fig. 2.13 for solders in the Sn–Ag–Bi system [27,28]. In solder wetting, the flux determines how much wetting actually occurs. The flux acts by reducing/limiting the amount of oxidation of the solder and substrate surfaces and creating a solder surface that approaches the intrinsic surface tension. Other factors in wetting include heat flow and fluid flow, including possible Marangoni flow.

Evaluation of solderability in manufacturing has considerably greater complexity than wetting of molten solder on a substrate in a controlled laboratory environment [29], but simple wetting balance and area-of-spread mea-

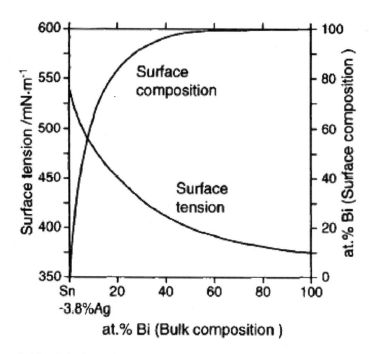

Figure 2.13   Calculated dependence of surface tension and surface composition as a function of bismuth composition for Sn–3.8Ag (atomic)–Bi alloys at 300°C. (From Ref. 27.)

surements are useful for separating the effects of some of these factors, and as screening tools when comparing different solder alloys. (Manufacturing issues are discussed in detail in Chapters 8 and 9.) Through numerous national and international lead-free solder R and D projects using wetting balance measurements, solderability was found to be a serious issue only for Zn-containing alloys, and then only for concentrations greater than 1% Zn.

For lead-free alloys not containing Zn, wetting characteristics on a specific metal substrate depend on the compositions of the solder and the substrate, the temperature of the solder and the substrate, the size and thermal conductivity of the substrate, the liquidus temperature of the solder, the surface condition of the substrate, the gaseous environment (oxygen, air, or nitrogen), and the flux. A comparison of wetting balance data for various lead-free solder alloys on copper from the IDEALS and NCMS projects indicates that 1) in general, the temperature for equivalent wetting balance performance to eutectic Sn–Pb scales with the liquidus temperature of the lead-free solder; and 2) the effects of the variables listed above are separable.

Fig. 2.12 from the IDEALS Lead-Free Project shows the time to 2/3 wetting force for five lead-free solder alloys compared with Sn–40Pb at three temperatures above the alloy liquidus temperature (T_1) per alloy, $T_1 + 25°C$, $T_1 + 35°C$, and $T_1 + 50°C$. With the exception of Sn–0.7Cu, the characteristic wetting times are virtually indistinguishable using Actiec 5 flux (Fig. 2.14a). When the flux is changed to pure Rosin flux, four of the five lead-free solders are again virtually identical to Sn–40Pb (Fig. 2.14b). Only Sn–0.7Cu–0.5Bi shows significantly poorer wetting than the other five solders.

Wetting balance data from the NCMS Lead-Free Solder Project exhibit the same trends and indicate, as well, that the effects of the experimental variables are separable. Fig. 2.15 shows an example of three wetting balance characteristic times for Sn–3.5Ag for two temperatures and five different fluxes. Wetting performance as indicated by these times varies systematically with flux and solder temperature.

2.3.2 Effect of Surface Finish on Wetting

The complexity quickly increases as manufacturing variables are included. In the NCMS Lead-Free Solder Project, the solderability of component leads was characterized using a semiquantitative "wetting figure-of-merit" as a function of solder composition, solder reflow temperature profile, and surface finish. The wetting performance of each alloy was evaluated during SMT assembly with pastes containing a conventional no-clean RMA (rosin mildly activated) flux and is illustrated in Fig. 2.16 as a function of the PWB surface finish. The wetting performance of the lead-free solders was almost as good as the eutectic Sn–Pb control, except when soldering to the imidazole OSP finish. At least one lead-free alloy matched the wetting performance of the eutectic Sn–Pb control alloy for each metal finish tested, other than the imidazole OSP-coated Cu. Most metallic surface finishes improved the spreading of the lead-free solders. In the case of the Ni/Au finish, all lead-free solders exhibited wetting scores of 5, the best performance possible. Immersion Sn finish also enhanced the spreading of the lead-free solders, most significantly in the case of the Sn–58Bi eutectic. On both the Ni/Pd and Pd-over-Cu finishes, the Sn-rich solders exhibited adequate wetting and spreading (equivalent to Sn or Ni/Au surface finishes), whereas Sn–58Bi and Sn–2.8Ag–20In exhibited considerably reduced spreading.

The IDEALS and NCMS results demonstrate that lead-free alloys can be differentiated, even ranked relative to each other, and to Sn–Pb eutectic. The issue is whether these differences are significant for manufacturing, performance, and reliability. As noted below, pull strength for leads in Fig. 2.18 did not exhibit any correlation with solder spreading and joint quality.

(a)

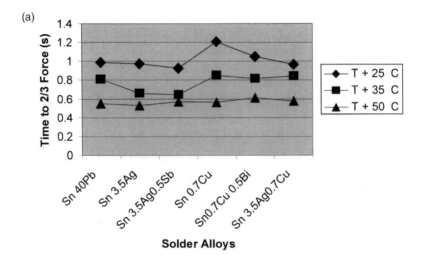

(b)

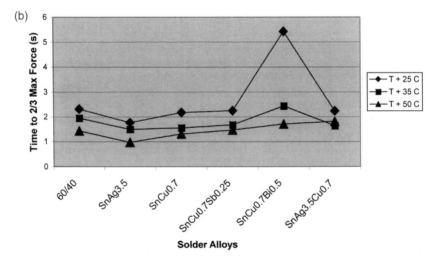

Figure 2.14 Time to 2/3 maximum force (in seconds) as a function of temperature above the liquids for five different Pb-free solder alloys as compared with Sn–40Rb. Wetting balance data for solder alloy wetting on copper (a) using Actiec 5 flux and (b) using pure rosin, unactivated (R) flux.

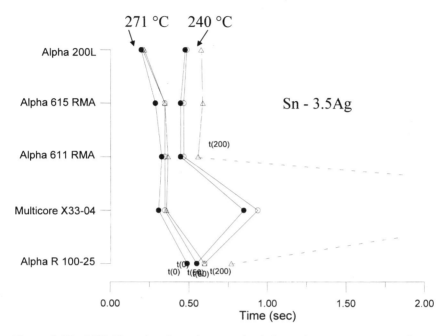

Figure 2.15 NCMS wetting data: three wetting balance time parameters as a function of flux type and temperature for Sn–3.5Ag.

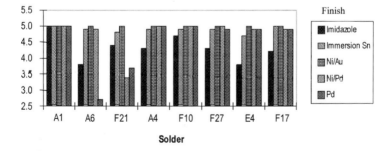

Figure 2.16 NCMS wetting figure of merit for SOIC device leads as a function of solder composition and surface finish.

2.4 RELIABILITY

2.4.1 Mechanical Properties of Lead-Free Solder Alloys

Solder alloys can be easily ranked based on ranges of mechanical property values allowable in a particular test or for a particular application. For solder alloys, however, one of the most important mechanical properties is resistance to thermomechanical fatigue (TMF). Unfortunately, the only method to determine TMF resistance is by doing TMF tests; a complicating factor is that for many materials, using methods and conditions that accelerate fatigue produces failure modes that are not relevant for real product applications.

For Sn–Pb solders, there is widespread acceptance of thermal cycling of printed circuit board test vehicles as a reliable method to assess TMF resistance. This acceptance comes from decades of industrial experience relating specific thermal cycling conditions to wear-out failure in product. From the NCMS Lead-Free Project, the lead-free solder alloys were able to withstand different amounts, types, and rates of loading which are dependent upon the different coefficients of thermal expansion (CTE) and mechanical properties of the board, components, and alloys, solder joint geometry, solder microstructure, and residual stresses. Taken together for a given alloy, these properties can produce solder joint performance better for some components than eutectic Sn–Pb and worse for other components on the same board, and may be different for different thermal cycling conditions [30,31].

In the NCMS Lead-Free Solder Project, the only surface mount components with obvious fatigue failures after more than 6700 cycles of 0°C to 100°C, or 5000 cycles of −55°C to +125°C were leadless ceramic chip carriers (LCCC) and 1206 chip resistors. No leaded surface mount devices exhibited failures. There were no unexpectedly early or catastrophic chip carrier or passive component failures. Those failures that occurred followed the same component order as observed for eutectic Sn–Pb. The ranking of alloys relative to eutectic Sn–Pb varied with thermal cycling conditions and component type.

The NCMS alloy ranking results, which change with component type, demonstrate the dangers of using a single component, a small subset of typical solder joint configurations, or a set of laboratory experiments, such as creep tests, to predict general behavior. The open questions, however, are what performance is necessary for lead-free solder alloys to be acceptable in most product applications and how well laboratory measurements and accelerated thermal cycling of test vehicles predict the performance of a given solder alloy relative to Sn–Pb eutectic. In this section, the thermal, compositional, and microstructure origins of the mechanical properties of lead-free alloys are examined to illustrate what mechanical behavior might be expected relative to Sn–Pb eutectic and other lead-free alloys.

The temperature and composition dependence of alloy mechanical properties is illustrated from the NCMS and IDEALS Project results using laboratory test methods. In the IDEALS Project, a range of physical properties of the lead-free solder alloys was measured, including coefficient of thermal expansion (CTE), elasticity, yield stress, and plastic behavior. Fig. 2.17 illustrates a significant point about lead-free solders as compared with Sn–Pb. The yield stress of near-eutectic Sn–40Pb is lower than for Sn–3.5Ag, Sn–0.7Cu–0.5Sb, and Sn–3.5Ag–0.7Cu for all temperatures. The solder Sn–0.7Cu exhibited the lowest yield stress at low temperatures but becomes virtually the same as the other lead-free alloys above 125°C. If the data are replotted using temperature normalized to their liquidus temperatures (homologous temperature), additional information can be obtained about their relative behavior. The yield stresses of Sn–40Pb, Sn–3.5Ag, Sn–0.7Cu–0.5Sb, and Sn–3.5Ag–0.7Cu are similar at low homologous temperature. At higher homologous temperatures, the yield stress of Sn–40Pb continues to decrease with increasing temperature, approaching zero at the eutectic temperature. For the lead-free solders, the yield stress shows a lower dependence on temperature as the homologous temperature increases. Maintaining strength with increasing temperature is characteristic of precipitation hardened materials. In the case of lead-free solders, the precipitation hardening is provided by the presence of the intermetallic phases, dispersed in and between the Sn dendrites. Creep results for lead-free solder alloys display similar transitions in behavior, leading to changing alloy rankings of creep resistance as the temperature and strain rate change.

2.4.2 Effect of Interaction Between Solder Alloys and Substrates on Joint Mechanical Properties

In the NCMS Lead-Free Solder Project, the interactions between solder alloys and surface finishes were characterized using mechanical pull testing of surface mount device leads for three different components on the manufacturing test vehicle. This test was used as a screening test to determine if any catastrophic failure modes arose with the lead-free alloys, and as a ranking test to differentiate among alloys. Approximately 100 device leads were pulled from each 256 I/O PQFP package and 10 to 15 device leads from each 20 I/O SOIC and 44 I/O PLCC package were pulled using a motorized microtensile tester at a crosshead speed of 0.015 cm/sec. The maximum load values ranged from 0.5 to 0.7 kg for the PQFP device leads, 1.1 to 1.8 kg for the PLCC device leads, and 1.6 to 3.2 lb for the SOIC device leads. The standard deviations for all measurements ranged from ±0.1 to ±0.3 kg. The maximum load values for SOIC device leads as shown in Fig. 2.18 are characteristic of the measurements for all three device types as a function of circuit board surface finish.

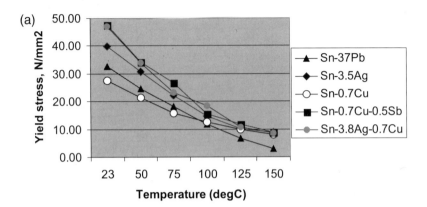

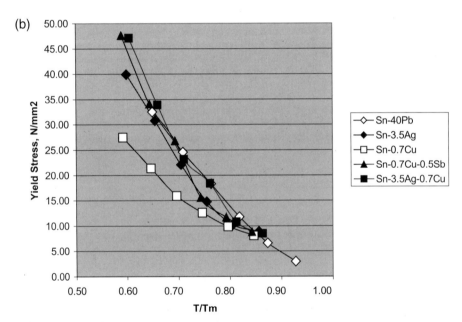

Figure 2.17 (a) Yield stress as a function of composition and temperature from IDEALS Lead-Free Project. (b) Yield stress as a function of composition and homologous temperature for Sn–40Pb and four lead-free solder alloys.

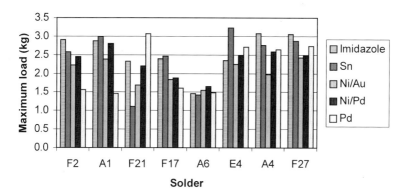

Figure 2.18 Pull strength of SOIC leads as a function of solder composition and surface finish.

Considering all board surface finish conditions, the only solder that showed statistically different pull strength than the other solders was Sn–58Bi. The maximum load values with Sn–58Bi are the lowest for imidazole, Ni/Au, Ni/Pd, and Pd, and second lowest for Sn. There was no correlation observed between the wetting figure of merit values, pull strength, and surface finish.

2.5 USING LABORATORY TEST RESULTS TO RANK LEAD-FREE SOLDER ALLOYS

The challenges in developing a set of laboratory-test-based criteria for accurately ranking lead-free solder alloys are illustrated by the ranking procedure used by the NCMS Lead-Free Solder Project. In the NCMS Lead-Free Solder Project, over 75 candidate alloys were identified at the start of the program as possible replacements for eutectic Sn–Pb. From this extensive list, the number of alloys had to be reduced before proceeding to full manufacturing and reliability trials. This "down-selection" process of appropriate lead-free solder alloys must involve trade-offs in properties "characteristic" of manufacturing and reliability. Does one choose an alloy, for example, that shows better wetting characteristics or another alloy that appears to have a better thermomechanical fatigue life? If the alloy with the better fatigue life is chosen, what is the risk of observing catastrophic problems with wetting during assembly? These choices were handled explicitly. First, pass–fail down-selection criteria were used to reduce the number of alloys on the list (Fig. 2.19). Second, the remaining alloys were grouped by alloy family, and only one alloy was selected based on the primary phase field in the phase

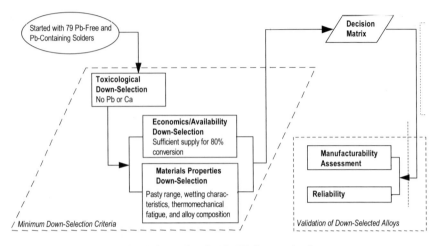

Figure 2.19   Schematic of NCMS down-selection process.

diagram. Last, a decision matrix was used to rank the remaining alloys, based on alloy pasty range, wetting balance values, and the results of a TMF test using a printed circuit board test vehicle. A full description of the decision matrix methodology, the test methods and how the decision matrix was applied in the NCMS Project can be found on the NCMS Lead-Free Project CD-ROM [1].

Table 2.2 shows the weighting factors and the scores for particular property values; property values between any two limiting values were interpolated. In this scheme, higher scores are better, with scores higher than eutectic Sn–Pb passing the down-selection process. With a weighting factor of

Table 2.2   NCMS Down-Selection Matrix, with Attribute Values and Property Weighting

NCMS Pb-free alloy down selection matrix

			Scale		
Test	Property	Weight	−10	0	5
DSC	Pasty range (°C)	10	30	5	0
Wetting	F_{max} (μN)	2	300	500	700
Balance	t_0 (sec)	2	0.6	0.3	0.1
	$t_{2/3}$ (sec)	2	1	0.45	0.5
TMF	TMF (% of Sn–37Pb)	10	75	100	150

10, the maximum score for a pasty range of $0\,°C$ is 50, while for a pasty range of $30\,°C$ it is -100. For the TMF test, a thermomechanical fatigue life greater than or equal to 150% of eutectic Sn–Pb earns a score of 50 points, while for a thermomechanical fatigue life of 75% of eutectic Sn–Pb, the solder would earn a score of -100.

Table 2.3a–c shows the results of the final decision matrix assessment for 10 lead-free solder alloys as compared with eutectic Sn–Pb. Table 2.3a–c presents the results based on only pasty range, on only pasty range and wetting, and on all three properties. Four alloys had negative scores based on pasty range alone. Considering both pasty range and wetting, these four alloys continued to have the lowest scores. All of the lead-free solder alloys tested had greater TMF lives than Sn–Pb eutectic. Considering the contribution of the thermomechanical fatigue test, these four alloys continued to have the lowest scores relative to Sn–Pb eutectic. (Note that two of these four alloys were not subjected to the TMF test; even if they had been tested and had achieved the maximum possible score for TMF, they would have still been on the list of four alloys with the lowest scores.) These decision matrix results indicate that the lowest ranked alloys were determined by their pasty ranges. Because only eight alloys, including Sn–Pb eutectic, could be carried through extensive manufacturing and reliability trials, three of four lowest ranked alloys were eliminated. (The alloy Sn–2.8Ag–20In was included in manufacturing and reliability trials because the NCMS Project members wanted to have one alloy with a liquidus temperature close to $183\,°C$ and, because it failed the decision matrix, it could be used as a test of the selection process.)

In the manufacturing and reliability trials, all seven lead-free alloys exhibited acceptable SMT manufacturability. The alloy Sn–2.8Ag–20In did indeed show the shortest fatigue life of any lead-free solder tested for both thermal cycling conditions and both components that showed solder joint

Table 2.3a NCMS Down-Selection Matrix for Candidate Pb-Free Solders: Pasty Range Scores Only

NCMS Pb-free solder project: decision matrix results												
Properties	Weight	A1	A4	A4	E3	E4	F2	F6	F10	F17	F21	F27
Pasty range (°C)	10	0	0	0	7	4	5	9	11	5	12	2
unweighted score		5	5	5	−0.8	1	0	−1.6	−2.4	0	−2.8	3
Weighted score before TMF score		50	50	50	−8	10	0	−16	−24	0	−28	30

Table 2.3b NCMS Down-Selection Matrix for Candidate Pb-Free Solders: Pasty Range and Wetting Scores Only

NCMS Pb-free solder project: decision matrix results

Properties	Weight	A1	A4	A4	E3	E4	F2	F6	F10	F17	F21	F27
Pasty range (°C)	10	0	0	0	7	4	5	9	11	5	12	2
unweighted score		5	5	5	-0.8	1	0	-1.6	-2.4	0	-2.8	3
F_{max} (μN)	2	278	337	248	298	295	329	325	283	315	333	418
unweighted score		-10	-8.15	-10	-10	-10	-8.55	-8.75	-10	-9.25	-8.4	-4.1
t_0 (sec)	2	0.33	0.33	0.433	0.28	0.23	0.23	0.26	0.33	0.28	0.76	0.59
unweighted score		-1	-1	-4.4	0.5	1.8	1.75	1	-1	0.5	-10	-9.7
$t_{2/3}$ (sec)	2	0.4	0.5	0.6	0.35	0.33	0.32	0.38	0.42	0.37	1.01	0.88
unweighted score		1.1	-0.73	-2.2	2	2.4	2.6	1.4	0.6	1.6	-10	-7.8
Weighted score before TMF score		30	30	17	-23	-2	-8	-29	-45	-14	-85	-13

Table 2.3c NCMS Down-Selection Matrix

NCMS Pb-free solder project: decision matrix results

Properties	Weight	A1	A4	A6	E3	E4	F2	F6	F10	F17	F21	F27
Pasty range (°C)	10	0	0	0	7	4	5	9	11	5	12	2
unweighted score		5	5	5	-0.8	1	0	-1.6	-2.4	0	-2.8	3
F_{max} (μN)	2	278	337	248	298	295	329	325	283	315	333	418
unweighted score		-10	-8.15	-10	-10	-10	-8.55	-8.75	-10	-9.25	-8.4	-4.1
t_0 (sec)	2	0.33	0.33	0.433	0.28	0.23	0.23	0.26	0.33	0.28	0.76	0.59
unweighted score		-1	-1	-4.4	0.5	1.8	1.75	1	-1	0.5	-10	-9.7
$t_{2/3}$ (sec)	2	0.4	0.5	0.6	0.35	0.33	0.32	0.38	0.42	0.37	1.01	0.88
unweighted score		1.1	-0.73	-2.2	2	2.4	2.6	1.4	0.6	1.6	-10	-7.8
TMF (% of Sn–37Pb)	10	100	164	200	200	191	164	N/A	155	155	N/A	155
unweighted score		0	5	5	5	5	5	-10	5	5	0	5
Total weighted score		30	80	67	27	48	42	-29	5	36	-85	37

Solder alloys:

A1 Sn–37 Pb
A4 Sn–3.5Ag
A6 Sn–58Bi
E3 Sn–3Ag–2Bi–2Sb
E4 Sn–3Ag–2Bi
F2 Sn–2.6Ag–0.8Cu–0.5Sb
F6 Sn–5Bi–7Zn
F10 Sn–0.2Ag–2Cu–0.8Sb
F17 Sn–4.8Bi–3.4Ag
F21 Sn–20In–2.8Ag
F27 Sn–3.5Ag–1Zn–0.5Cu.

failures. For the remaining alloys, the TMF life and, hence, the relative alloy ranking depended on thermal cycling conditions and component type.

2.6 SUMMARY

The behavior of solder alloys in manufacturing and in use can be understood in terms of their thermodynamic properties, the kinetics of reactions, including wetting, and their temperature, stress, and strain-rate-dependent mechanical properties. While the performance of a specific solder alloy cannot be quantitatively predicted in manufacturing or in product applications based on laboratory experiments, the metallurgical concepts outlined in this chapter combined with laboratory measurements can be used to identify alloys that have potential for commercial use. The thermodynamics of alloy melting, solidification, and wetting forms the basis for understanding solder joint formation in reflow and wave soldering applications. Once a printed circuit board is successfully assembled, the reliability of its solder joints in use depends not only on the solder's thermomechanical properties, including thermal expansion coefficient and the response of the alloy microstructure to the applied stress, but also on the properties and the response of the components and the circuit board in the system. By examining the dependence of the mechanical properties of lead-free and Sn–Pb eutectic solder as a function of temperature, strain-rate, and stress, the underlying mechanisms responsible for solder behavior changing as a function of alloy composition are beginning to be revealed.

REFERENCES

1. NCMS Lead-Free Solder Project Final Report. Report 0401RE96, National Center for Manufacturing Sciences, 3025 Boardwalk, Ann Arbor, Michigan 48108-3266, 1997, and CD-ROM database of complete dataset, including micrographs and raw data, August, 1999. Information on how to order these can be obtained from: http://www.ncms.org/.
2. I Artaki, D Noctor, C Desantis, W. Desaulnier, L Felton, M Palmer, J Felty, J Greaves, CA Handwerker, J Mather, S Schroeder, D Napp, TY Pan, J Rosser, P Vianco, G Whitten, Y Zhu. Research trends in lead-free soldering in the US: NCMS lead-free solder project (Keynote). IEEE Computer Society Proceedings—EcoDesign '99: First International Symposium on Environmentally Conscious Design and Inverse Manufacturing. Tokyo, 1999, pp 602–605.
3. CA Handwerker, EE de Kluizenaar, K Suganuma, FW Gayle. Major international lead (Pb)-free solder studies. In: KJ Puttlitz, KA Stalter, eds. Handbook

of Lead-Free Solder Technology for Microelectronic Assemblies. New York: Dekker, 2004.

4. CA Handwerker. NCMS lead-free solder project: a summary of results, conclusions and recommendations. IPC. IPC Work '99: An International Summit on Lead-Free Electronics Assemblies; Proceedings; October 23–28, 1999; Minneapolis, MN.

5. M Harrison, JH Vincent. Improved Design Life and Environmentally Aware Manufacturing of Electronic Assemblies by Lead-Free Soldering. < *http://www.lead-free.org/research/index.html* >.

6. MR Harrison, J Vincent. IDEALS: Improved design life and environmentally aware manufacturing of electronics assemblies by lead-free soldering. Proc. IMAPS Europe'99, (Harrogate, GB), June 1999.

7. The Synthesis Report for the IDEALS project can be downloaded from: < *http://www.alphametals.com/products/lead_free/PDF/synthesis.pdf* > and < *http://www.cordis.lu* >.

8. MH Biglari, M Oddy, MA Oud, P Davis, EE de Kluizenaar, P Langeveld, D Schwarzbach. Lead-free solders based on SnAgCu, SnAgBi, SnAg, and SnCu, for wave soldering of electronic assemblies. Proc. Electronics Goes Green 2000+, (Berlin, Germany), September 2000.

9. J Bath, CA Handwerker, E Bradley. Research Update: Lead-Free Solder Alternatives. Circuits Assembly, May 31–40, 2000.

10. E Bradley. NEMI lead-free interconnect task group report. IPC Work '99: An International Summit on Lead-Free Electronics Assemblies. Proceedings; ICP. Minneapolis, MN, 1999.

11. K Suganuma. Research and development for lead-free soldering in Japan. IPC Work '99: An International Summit on Lead-Free Electronics Assemblies; Proceedings; IPC. October 23–28, 1999; Minneapolis, MN.

12. NEDO Research and Development on Lead-Free Soldering. Report No. 00-ki-17, JEIDA, Tokyo, 2000.

13. Project symposium on research and development on lead-free soldering for standardization. JWES, Tokyo, Japan, 2000.

14. KW Moon, WJ Boettinger, UR Kattner, FS Biancaniello, CA Handwerker. Experimental and thermodynamic assessment of Sn–Ag–Cu solder alloys. J Electron Mater 29(10):1122–1136, 2000.

15. UR Kattner, WJ Boettinger. On the Sn–Bi–Ag ternary phase-diagram. J Electron Mater 23:603–610, 1994.

16. UR Kattner, CA Handwerker. Calculation of phase equilibria in candidate solder alloys. Z Metall 92(7):740–746, 2001.

17. UR Kattner. Phase diagrams for lead-free solder alloys. J Min Met Mats 54(12): 45–51, 2002.

18. KW Moon, WJ Boettinger, CA Handwerker, DJ Lee. The effect of Pb contamination on the solidification behavior of Sn–Bi solders. J Electron Mats 30(1):45–52, 2001.

19. Y Kariya, N Williams, C Gagg, W Plumbridge. Tin pest in Sn–0.5 wt.% Cu lead-free solder. JOM—J Miner Met Mater Soc 53, 39–41, 2001.

20. HM Lee, SW Yoon, BJ Lee. Thermodynamic prediction of interface phases at Cu/solder joints. J Electron Mater 27:1161–1166, 1998.

21. JH Vincent, BP Richards, DR Wallis, I Gunter, M Warwick, HAH Steen, PG Harris, MA Whitmore, S Billington, AC Harman, E Knight. Alternative solders for electronics assemblies: Part 2. UK progress and preliminary trials. Circuit World 9:32–34, 1993.

22. Alternative Solders for Electronic Assemblies-Final Report of DTI Project 1991–1993. GEC Marconi, ITRI, BNR Europe, and Multicore Solders. DTI Report MS/20073, issued 10.26.93.

23. JH Vincent, G Humpston. Lead-free solders for electronic assembly. GEC J Res 11:76–89, 1994.

24. WJ Boettinger, CA Handwerker, B Newbury, TY Pan, JM Nicholson. Mechanism of fillet lifting in Sn–Bi alloys. J Electron Mater 31(5):545–550, 2002.

25. K Suganuma. Microstructural features of lift-off phenomenon in through hole circuit soldered by Sn–Bi. Scr Mater 38(9):1333–1340, 1998.

26. H Takao, H Hasegawa. Influence of alloy composition on fillet-lifting phenomenon in Sn–Ag–Bi alloys. J Electron Mater 30:513–520, 2001.

27. I Ohnuma, XJ Liu, H Ohtani, K Anzai, R Kainuma, K Ishida. Development of thermodynamic database for micro-soldering alloys. Proceedings of 3rd Electronics Packaging Technology Conference (EPTC 2000), 2000, pp 91–96.

28. I Ohnuma, M Miyashita, K Anzai, XJ Liu, H Ohtani, R Kainuma. Phase equilibria and the related properties of Sn–Ag–Cu based lead-free solder alloys. J Electron Mater 29(10):1137–1144, 2000.

29. N-C Lee. Prospect of lead-free alternatives for reflow soldering. IPC Work '99: An International Summit on Lead-Free Electronics Assemblies; October 23–28, 1999; Minneapolis, MN.

30. G Swan, A Woosley, K Simmons, T Koschmieder, TT Chong, L Matsushita. Development of lead-(PB) and halogen free peripheral leaded and PBGA components to meet MSL3 at 260 C peak reflow profile. Proceedings, Electronics Goes Green 2000+, Berlin: VDE Verlag, 2000, pp 121–126.

31. J Bartelo, SR Cain, D Caletka, K Darbha, T Gosselin, DW Henderson, D King, K Knadle, A Sarkhel, G Thiel, C Woychik. Thermomechanical fatigue behavior of selected lead-free solders. Proceedings IPC SMEMA Council APEX 2001, Paper # LF2-2.

3
Alloy Selections

Kay Nimmo

Soldertec Global at Tin Technology Ltd., St. Albans, Hertfordshire, United Kingdom

To make an informed choice of alloy from the range of lead-free solders available, it is necessary to understand a range of factors; not all of them related to scientific or process issues, but some to commercial matters such as cost and patent/license rights. It is also very important to note that the choice of lead-free solder is more application specific than has been the case in the past with Sn–Pb; the process method, product type and design, and the resulting constraints can each have a significant effect. Achieving a suitable balance between all these issues makes the *application-specific lead-free alloy choice* successful.

3.1 ALLOY POSSIBILITIES AND REQUIREMENTS

A huge number of lead-free alloy compositions can be imagined from combinations of elements: Sn, Ag, Bi, Cu, In, Sb, Zn in binary, ternary, or multi-component systems. Other minor additions can also be made from elements such as Au, Fe, Ge, La, Mg, Ni, Pd, and many others, for purposes of grain refining or other strengthening mechanisms. Indeed, many hundred patent applications have been filed around the world intending to cover specific combinations of alloy ranges [1]. However, when it comes to selection of a lead-free solder many of these alloy combinations can be discounted as

49

adding compositional control difficulties with limited practical benefits for the soldering process or the final reliability of the soldered joint. Exceptions may be elements added to improve sphericity of solder balls or other specific purposes. The important lead-free alloys are identified and described in this chapter. The major constituent is denoted first, followed by alloying elements in order of decreasing percentage. It should be remembered that only generalized comparisons between alloys are possible in a short chapter such as this and many other details are available in literature that can be consulted in order to obtain additional precise data.

All alloys, even the traditional Sn–Pb eutectic, have various advantages and disadvantages, and various properties that need to be considered when making such a selection for any particular product. Some of these factors are summarized below:

- Physical properties (melting temperature, surface tension, electrical and thermal conductivity, specific heat capacity, expansion, and shrinkage characteristics, etc.).
- Process issues (wettability, dissolution rates of substrate and pot materials, flux interaction and shelf life, drossing behavior, intermetallic formation, formability into wire or performs, etc.).
- Mechanical properties (tensile strength and ductility, creep resistance, shear strength, fatigue resistance, crack propagation rate, microstructural aging, hardness, resistance to strength reduction from impurity contamination, etc.).
- Other influences on service performance (corrosion resistance, embrittlement, intermetallic formation and distribution, etc.).
- Commercial factors (metal cost, ease of powder production or wire formation, development time, license fees and permits, etc.).
- Environmental acceptance (recyclability, low toxic potential, etc.).

Data are available on these properties for Sn–Pb and most of the major lead-free systems. Some will be found in this book, but other good sources include solder alloy suppliers, industry research organizations, and literature review papers [2–15].

3.2 ALLOY MELTING TEMPERATURE AND THE SOLDERING PROCESS

The effects of the solder melting temperature, and how well it will fit into the production process, are perhaps the first factor to consider when selecting an alloy. The maximum temperature of any production process is limited by the thermal properties of components on the board, while the minimum is defined by the melting behavior of the solder and the requirement to form a good joint with the substrate. These limitations are outlined in Fig. 3.1 which demonstrates the slightly limited reflow process window available with most lead-free alloys when compared with Sn–Pb. In summary, the window lies between

- Highest endurable temperature, and shortest time, to prevent moisture related pop-corning of packages, degradation of polymeric materials, fluxes, etc.
- Lowest temperature that will ensure that the solder melts, forms a joint with the substrate, and re-solidifies within the shortest possible process time.

It is these limitations that often lead to the addition of third, fourth, or even more alloying elements in attempts to lower the melting point by even a few degrees. In most cases, the added complication of highly engineered alloys, with more than three elements, does not provide sufficient property benefits for their use, and alloys other than those described in this chapter

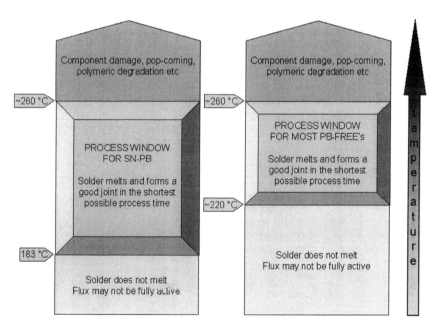

Figure 3.1 Process window restrictions with lead-free solders. The more limited process window for most lead-free alloys requires improved process control, production equipment, and appropriate alloy selection. Process window restrictions are exaggerated with fast production processes or presence of large components. Paste may exhibit flux deterioration with slow profiles.

should rarely be required. Exceptionally, however, the addition of low levels of Bi is effective in achieving a lowering of melting point while maintaining, or even improving, desired service life, for example, in a Sn–Ag–Cu–Bi solder. In another example, the addition of trace levels of Ni has also been found to be of benefit to the Sn–Cu binary system.

Production equipment and the entire basis of assembly design has been optimized for the use of Sn–Pb eutectic over the last 30 years and it is therefore important to consider to what stage lead-free implementation within a process has been achieved. An alloy selected for use with older style production equipment where good temperature control may be problematic, and, where some possibility of Pb contamination exists, may not be the optimum choice, even for an identical assembly, if the process has been initially designed to utilize energy-saving lead-free compatible production equipment, suitable components and other materials. Both processability and product reliability can be affected by process conditions and/or the presence of Pb. Examples of phases that can form through contamination with Pb, and importantly their melting temperature, can be found in Table 3.1. The greater the pasty range, i.e., the greater the difference in formation temperatures of different phases in an alloy, the greater is the risk of fillet lifting or hot tearing defects.

Table 3.1 Potential "Low
Temperature Phases" in
Combination with Pb

Alloy (wt.%)	°C
Ternary	
63Sn–37Pb–0.1Cu	182
63Sn–36Pb–1Ag	177
Sn–Pb–Zn	176
56Sn–40Pb–4Sb	175
52Bi–30Pb–18Sn	95
Binary	
63Sn–37Pb	183
58Bi–42Sn	138
56Bi–44Pb	124

Source: Calculated by AT Dinsdale, National Physical Laboratory, U.K., using MTDATA and NPL Solders Database, August 2002. Systems have not been fully assessed but provide a useful indication. Compositions in particular are not precise.

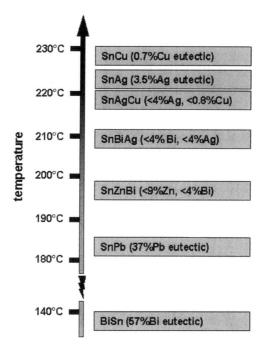

Figure 3.2 Schematic diagram of lead-free solder melting temperatures.

A rough generalized comparison of solder melting temperature can be seen in Fig. 3.2. Eutectic alloys have precise melting points similar to pure metals, but noneutectic solders melt over a temperature range between the solidus temperature (at which the alloy is completely solid) and the liquidus temperature (at which the alloy is completely liquid). Consideration of this factor is also more important in the reflow process, where the assembly must be held above melting point for perhaps 30–90 sec, than the wave solder process where contact with the solder bath is limited to only a few seconds. It is for this reason that the alloys found toward the top of Fig. 3.2 (Sn–Cu) can be used and are favored by some, for wave soldering, but are generally less often used in the reflow process.

3.3 ALLOY MELTING TEMPERATURE AND MECHANICAL PROPERTIES

It is not only the soldering process and resulting microstructure that needs to be understood, but also the potential relation of actual melting temperature of any alloy to mechanical properties, particularly at high temperature. It has often been noted that solder alloys are required to work under more extreme service conditions than some of the highly developed nickel superalloys designed for aerospace use. This view arises from examination of the comparative homologous temperatures of engineering superalloys and normal solders, i.e., the temperature of test or service divided by alloy melting point (both in Kelvin). This homologous temperature provides an idea of how severe the service temperature actually is and, for instance, the likelihood of creep occurring (significant above 0.5). At room temperature, the homologous temperature for Sn–Pb is 0.65 and for Sn–Ag is 0.60, while at 100°C, Sn–Pb is working at 0.82 and Sn–Ag at 0.76. All solder alloys are therefore expected to exhibit significant creep even at room temperature, but, partially

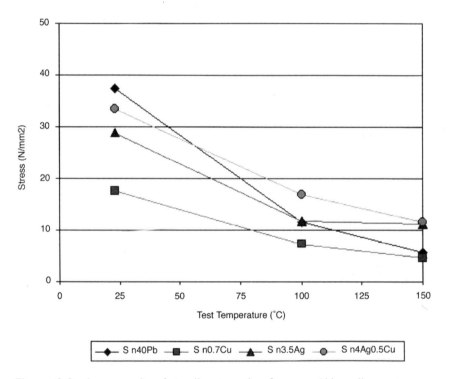

Figure 3.3 An example of tensile strength of some solder alloys at room temperature and above. Tests carried at Soldertec on aged, cast samples, of 2 mm diameter, at 5×10^{-3}/sec. Other strain rates produced a similar comparison between alloy behavior at temperature. Other sample and test conditions may produce a different result, for example, lower strength for Sn–Pb when compared with the lead-free solders. (From Ref. 33.)

as a result of its "low" melting point, and partially due to microstructure, Sn–Pb typically has an inherently lower creep strength when compared to lead-free solders, particularly as temperature increases further.

The use of this homologous temperature concept often leads to the expectation that lead-free alloys will outperform Sn–Pb in all mechanical properties, including fatigue. This may typically be true for basic mechanical properties, for example, Sn–37Pb loses almost 70% of its tensile strength as the test temperature is raised from 20 to 100°C, whereas, in comparison, the Sn–4Ag–0.5Cu alloy loses only around 50% over the same temperature range. This is demonstrated in Fig. 3.3, which provides an indication of relative alloy strength; however, it should be remembered that other test conditions could result in lower strength for Sn–Pb when compared with the lead-free solders. In addition, in practical soldering situations, the situation is more complex and other factors of significance also come into effect when considering fatigue life, process conditions, microstructure, substrate reactions, crack growth rates, expansion coefficients, etc. Some of the complex interactions of factors affecting fatigue life are illustrated in Fig. 3.4, and examples of the variety of intermetallic phases possible in the bulk of solder alloys, and at Cu or Ni interfaces, each with varying potential effect, can be seen in Table 3.2. Many studies report on this topic.

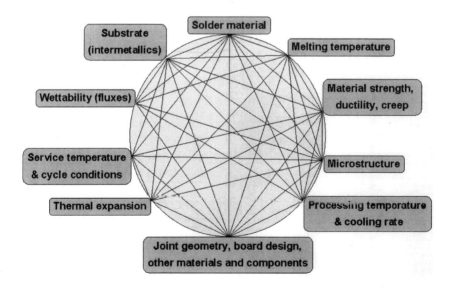

Figure 3.4 Some of the complex factors affecting fatigue life.

Table 3.2 Potential Intermetallic Compounds

Alloy	In bulk solder	Cu substrate	Ni substrate
Sn–3.5Ag	Ag_3Sn	Cu_6Sn_5, Cu_3Sn	Ni_3Sn_4, Ni_3Sn, Ni_3Sn_2, $NiSn_3$
Sn–3.4Ag–0.8Cu	Ag_3Sn, Cu_6Sn_5, Cu_3Sn	Cu_6Sn_5, Cu_3Sn	$(Ni,Cu)_3Sn_4$, $(Ni,Cu)_6Sn_5$
Sn–3Ag–1Bi–0.5Cu	Ag_3Sn, Cu_6Sn_5, Cu_3Sn	Cu_6Sn_5, Cu_3Sn	$(Ni,Cu)_3Sn_4$, $(Ni,Cu)_6Sn_5$, $NiBi_3$, NiBi
Sn–3Ag–3Bi	Ag_3Sn	Cu_6Sn_5, Cu_3Sn	Ni_3Sn_4, Ni_3Sn, Ni_3Sn_2, $NiSn_3$, $NiBi_3$, NiBi
Sn–0.7Cu	Cu_6Sn_5, Cu_3Sn	Cu_6Sn_5, Cu_3Sn	$(Ni,Cu)_3Sn_4$, $(Ni,Cu)_6Sn_5$
Sn–9Zn	–	CuZn, Cu_5Zn_8, Cu_6Sn_5, Cu_3Sn	Ni_3Sn_4, Ni_3Sn, Ni_3Sn_2, $NiSn_3$
Sn–8Zn–3Bi	–	CuZn, Cu_5Zn_8, Cu_6Sn_5, Cu_3Sn	Ni_3Sn_4, Ni_3Sn, Ni_3Sn_2, $NiSn_3$, $NiBi_3$, NiBi
Bi–43Sn	–	Cu_6Sn_5, Cu_3Sn, Cu–Sn–Bi	Ni_3Sn_4, Ni_3Sn, Ni_3Sn_2, $NiSn_3$, $NiBi_3$, NiBi
Bi–42Sn–0.5Ag	Ag_3Sn	Cu_6Sn_5, Cu_3Sn, Cu–Sn–Bi	Ni_3Sn_4, Ni_3Sn, Ni_3Sn_2, $NiSn_3$, $NiBi_3$, NiBi
Sn–5Sb	Sb_2Sn_3	Cu_6Sn_5, Cu_3Sn	Ni_3Sn_4, Ni_3Sn, Ni_3Sn_2, $NiSn_3$
In–48Sn	–	Cu_6Sn_5, Cu_3Sn, Cu–In–Sn and Cu–In e.g. $Cu_{37}In_{53}$	In–Ni e.g. $In(Sn)_3Ni_2$, Ni_3Sn_4, Ni_3Sn, Ni_3Sn_2, $NiSn_3$

Compounds potentially formed in solder systems—variable in nonequilibrium systems (different cooling rates) and on substrate materials. Different compounds may be seen on a plated coating rather than on a bulk substrate of the same material.

Notes: $NiSn_3$ more likely on plated layers [16]. $NiBi_3$ found but not NiBi [17]. No Ni phase seen with Ag, Bi, Sb [18]. In_3Ni_2 formed [19]. See also Ref. 15, Ch. 20.

It should also be remembered that no direct correlation has yet been established between basic mechanical properties of solders, such as tensile strength, and creep strength of bulk solders or solder joints, and fatigue service life. This is illustrated in Fig. 3.5 which shows significant variations in creep, shear, and tensile strength between three alloys, Sn–37Pb, Sn–36Pb–2Ag, and Sn–3.5Ag, but almost identical fatigue life [12]. Although in this case fatigue was measured at one temperature (isothermal) under applied stress and is therefore not identical to thermally cycled, strain-controlled conditions experienced in solder joints, it provides a useful example. As another example, the Bi–Sn eutectic alloy is reported to have excellent fatigue life at conditions very close to its melting point (up to 125 °C) where tensile strength is relatively low [20].

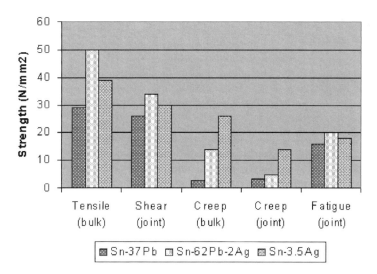

Figure 3.5 Comparison of mechanical and fatigue test data. Differences in strength do not directly relate to fatigue performance. Testing at 20°C and 0.2 mm/min. Stress controlled creep and fatigue life at 1000 hr, specimens aged at room temperature. (Data from Ref. 12, graph format from H Steen, Loctite.)

When carrying out mechanical tests or examining test results, it is very important to note the exact sample preparation route, including thermal history, and the age at which the mechanical measurement has been made, including any heat treatment. If these details are not known the data obtained are practically useless because the results depend very significantly on the initial microstructure and stability over time.

The performance of Sn–Pb is well accepted despite being far from ideal in some applications, and, overall, it is possible to obtain equivalent or superior fatigue performance to the currently used Sn–Pb alloy with the use of the correct lead-free solder. *It is important to assess the suitability of each alloy for the situation in which it will be applied.*

3.4 MAJOR LEAD-FREE ALLOY TYPES

3.4.1 Sn–Ag

The Sn–Ag eutectic has been used in many applications, over many years, not because of its "lead-free" properties but because it provides good fatigue resistance and was found to be particularly suited to some of the most demanding applications. Sn–Ag has also been used as a slightly higher temperature alloy than the Sn Pb eutectic, as an intermediate between Sn–Pb and Pb–Sn (high-temperature solders) with a history of reliability in such situations.

The binary eutectic exists at around 3.5% Ag and a temperature of approximately 221°C. The microstructure is a eutectic of a Sn-rich phase with very low levels of Ag in solid solution and fine Ag_3Sn intermetallics that tend to form as thin platelets. The Ag_3Sn particles are difficult to nucleate and the eutectic composition often shows a structure consisting of primary Sn dendrites in a divorced eutectic due to this factor. This may be particularly noticeable at high solidification rates where nucleation is further suppressed and a greater percentage of Sn-rich dendrites are seen (see Fig. 3.6) [21]. Microstructural development toward a more stable structure closer to the equilibrium state during aging produces changes in alloy properties, generally with a reduction in strength and some increase in ductility.

With compositions of higher Ag content than the eutectic alloy there is a very steep rise in liquidus temperature, for example, to around 250°C at 5% Ag. These high temperatures effectively limit the maximum Ag content of alloys that can be of practical use. Alloys of 5% Ag may be used for some specialist applications, the structure of which consists of primary Ag_3Sn

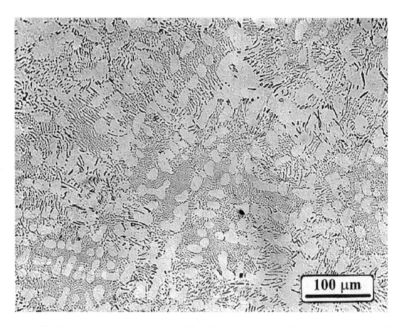

Figure 3.6 Typical microstructure of Sn–3.5Ag. Primary Sn-rich dendrites (light) and eutectic regions. (From J.S. Hwang. Environment-friendly electronics: Lead-free technology. Electrochemical Publications Ltd., Isle of Man, British Isles, 2001.)

needles in a eutectic Sn–Ag_3Sn eutectic matrix; however, most electronic solders will consist of 4% Ag or lower.

Because the solid solubility of Ag in Sn is low (below 0.1%) there is little solid solution hardening achieved by alloying, and the good strength of the solder is improved by the fine dispersion of intermetallic particles, which have also been reported to limit crack propagation rates and improve fatigue life [14]. As with many other Sn-based solders, the bulk tensile strength can be lower than that of Sn–Pb at room temperature but higher in relative terms, at temperatures above around 100°C, as the strength of Sn–Pb reduces more dramatically with temperature (example shown in Fig. 3.3). Studies of the fatigue performance of Sn–Ag normally demonstrate equivalence or superiority to Sn–Pb, although, as with all lead-free alloys, in some instances Sn–Pb may be superior [21,22]. For example, while Sn–3.5Ag may be superior to Sn–Pb at plastic strains of <10%, this appears to be no longer the case at low strains at high temperature (150°C) [23].

The microstructure can also be affected by the substrate material and, for example, sufficient Cu can dissolve in the Sn–Ag binary during the soldering process for Cu_6Sn_5 intermetallic particles to form in the bulk of the solder, as well as at the interface. In this way, the microstructure can be similar to that of the Sn–Ag–Cu ternary solder described in Sec. 4.2 when used with Cu substrate. Fig. 3.7 shows the effects of Cu dissolution on the microstructure of Sn–3.5Ag solder and, specifically, the formation of Cu_6Sn_5 intermetallic compounds in the bulk of the joint [24,25]. Following aging, fewer Cu_6Sn_5 particles will be expected in the solder immediately adjacent to the Cu substrate itself as Cu from this region will be depleted through formation of the interfacial intermetallic layer itself, and the mechanical properties will also be subtly different.

It is possible to form a low melting temperature phase in Sn–Ag joints through contamination with Pb (see Table 3.1). The ternary eutectic melts at 179°C, tends to concentrate in ternary phase pockets at grain boundaries, and may lead to associated process problems (such as fillet lifting), or joint strength reduction, specifically at higher service temperatures [26]. However, as the temperature of the Pb-containing phase (179°C) is not as far from the original melting point of the solder itself (221°C) as some Bi phases, the detrimental effects are not as great as those expected through Pb contamination of Bi containing solders.

It should also be noted that there is no evidence of "Ag migration" from this, or other Ag containing lead-free solders. The Ag in these alloys is mainly in the form of Ag_3Sn intermetallic, rather than in solid solution, and is reported to be stable against dissolution or dissociation by corrosive environments [27].

The binary Sn–3.5Ag solder is not subject to any patent restrictions [1].

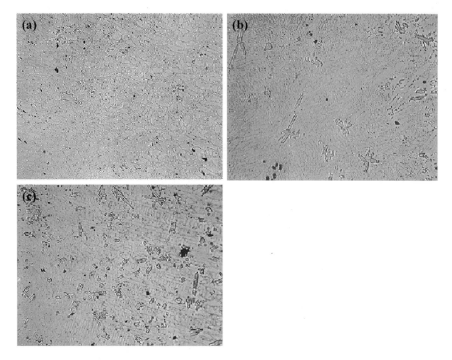

Figure 3.7 Microstructure of Sn–3.5Ag solder affected by Cu dissolution from a substrate. Cu_6Sn_5 intermetallics can be clearly seen in the bulk of the alloy, increase in number and size with peak reflow time (as shown here), and also with time above melting point. Reflow at (a) 236°C, (b) 269°C, (c) 292°C. (From J.S. Hwang. Environment-friendly electronics: Lead-free technology. Electrochemical Publications Ltd., Isle of Man, British Isles, 2001.)

3.4.2 Sn–Ag–Cu–(Bi)

Low levels of copper addition to the Sn–Ag binary system can have the following beneficial effects:

- Lowering of the melting point, with the possible formation of a ternary eutectic.
- Improvement in wettability.
- Improvement of the thermal–mechanical performance of the alloy under certain service conditions.
- Retardation of the dissolution of copper from substrates.

This slight improvement in several properties over those of the binary Sn–Ag alloy, and the fact that it is a Bi-free solder less affected by Pb contamination, has led to selection of Sn–Ag–Cu as the currently most favored lead-free solder.

Use of the ternary eutectic composition ensures that the minimum melting temperature and a fine, homogeneous dispersion of intermetallic compounds will be achieved with resulting good and stable mechanical properties and fatigue life. Various studies have therefore attempted to locate the precise ternary eutectic composition for the Sn–Ag–Cu solder, and while it is generally agreed that the ternary eutectic temperature is close to 217°C there is more discussion over the precise composition. Although one study proposes Sn–4.7Ag–1.7Cu as the ternary eutectic point [28] this seems unlikely. This composition is reported to be well outside the region of the phase diagram that can reasonably be expected to exhibit eutectic behavior, and most experimental studies, and thermodynamic phase diagram calculations, point to an area of the phase diagram with significantly lower Ag and Cu contents in the region of Sn–3.4Ag–0.8Cu [29–31]. Provided that the solder composition used is in the region of this approximate eutectic point (around 3.3–3.7Ag and 0.7–0.9Cu), joint properties can be expected to be similar when formed under similar process conditions; the exact eutectic is not required in practice. Of course, in practical application, precise microstructural control cannot be expected, not just because of compositional variations but also because of differences in processing technique, temperature, and dissolution rate of substrates which, in combination, can have significant effects.

The true eutectic will exhibit a fine microstructure consisting of solid solution Sn, Cu_6Sn_5, and Ag_3Sn intermetallics; a typical microstructure is shown in Fig. 3.8. Excess Cu will tend to promote the formation of primary

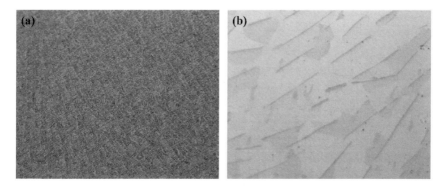

Figure 3.8 Microstructure of Sn–3.4Ag–0.8Cu near eutectic. Magnification of (a) ×200 and (b) ×1000. (From Ref. 29.)

Cu_6Sn_5 needles, and excess Ag will tend to promote formation of large primary Ag_3Sn needles. Dissolution of either of these elements (Cu or Ag) from external sources such as board or component coatings can alter the composition of the final joint and it may therefore be preferable, for example, to use a lower Ag content alloy with an Ag board finish in order to maintain the expected microstructure and properties. The driving force for dissolution of Cu from substrate materials is lower if Cu is already present in the system. Dissolution into the ternary Sn–Ag–Cu is therefore slower than into the binary Sn–Ag with associated benefits of easier solder bath composition control and reduced effect on the circuit board itself. However, as with the Sn–3.5Ag eutectic alloy, primary Cu_6Sn_5 intermetallics may be seen in the bulk of the solder joint as the Cu level is increased. Differences will also be noted in the microstructure formed near, for example, a Cu substrate where the Cu will be depleted from the matrix in the interfacial region close to the area of Cu–Sn intermetallic formation, and will thus limit growth of the Cu_6Sn_5 intermetallic phase in that area of the bulk of the joint during aging.

Use of the Sn–4.7Ag–1.7Cu alloy will therefore be more prone to formation of microstructures containing large primary Cu_6Sn_5 intermetallic compounds with potentially embrittling effects than alloys closer to the approximate eutectic. Creep and fatigue performance may, for example, be compromised, particularly if the primary intermetallics are found in areas of stress concentration, such as through preferential formation at the interface with a Cu substrate. Primary intermetallic compounds can also be found in solder alloys close to, but not on, the eutectic composition. However, this is only likely to occur in situations where cooling rates are slow enough to allow precipitation, a situation that is unusual in normal reflow profiles. It should also be remembered that the fine intermetallics formed in the eutectic phase are generally beneficial, strengthening the joint and also potentially reducing crack propagation rates. This may enhance the fatigue lifetime of the solder if the optimum, fine eutectic microstructure can be achieved during processing. Such a microstructure is expected to be more stable during aging, potentially providing more predictable fatigue properties. Fig. 3.9 shows the differences between microstructures of a noneutectic and an approximate eutectic solder during thermal cycling. Through comparison of mechanical property data comparing various Sn–Ag–Cu solders tested under approximately similar conditions, it would appear that the high alloy Sn–4.7Ag–1.7Cu solder has the highest strength but lowest ductility, whereas in reverse, the low Ag alloy Sn–3Ag–0.5Cu solder has the highest ductility but lowest strength [32,33].

It is possible to form a low melting temperature phase in Sn–Ag–Cu joints through contamination with Pb through formation of a eutectic melting at 179°C [26], but it has been thermodynamically calculated that even in a Sn–Ag–Cu system containing up to around 3.5% Pb solidification will be

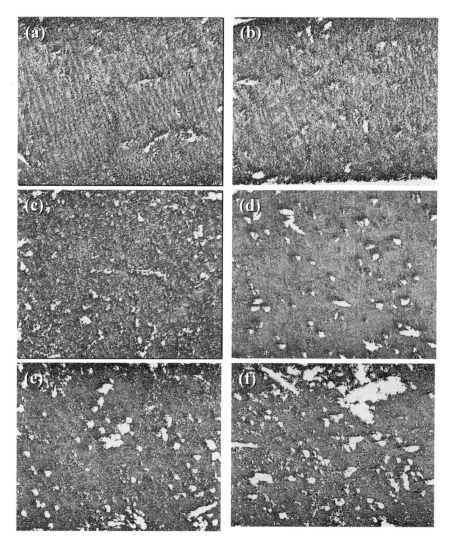

Figure 3.9 Microstructure of Sn–3.4Ag–0.8Cu and Sn–4.7Ag–1.7Cu after thermal cycling. Illustrating the increasing intermetallic fraction in Sn–3.4Ag–0.8Cu during cycling over (a) 300, (b) 600, and (c) 900 cycles of −35 to +85°C, and how this is much more evident in the Sn–4.7Ag–1.7Cu solder after the same conditions over (d) 300, (e) 600, and (f) 900. Magnification ×100. (From Ref. 29.)

completed by the eutectic temperature of 217°C. However, even below this level the Pb-rich phase has a tendency to gather around the grain boundaries to the detriment of the mechanical properties of the solder [29,34]. Therefore although Sn–Ag–Cu is not as prone to Pb contamination problems as Bi containing solders, contamination with Pb should still be avoided wherever possible.

It should be noted that some compositions are subject to patent/license restrictions in some countries, but solder suppliers should be aware of, and be able to provide information on, this issue. Some concerns regarding conflicting patents on Sn–Ag–Cu are being addressed through cross-licensing agreements. Some consumers may prefer to focus on a technically acceptable composition (Sn–4Ag–0.5Cu) that is thought to be patent-free due to its publication in 1959 [1,35].

The addition of Bi to the ternary alloy has been used in order to provide additional solid solution strengthening and potentially improve fatigue performance with the possibility of reducing Ag content and therefore cost. As a result of solid solution strengthening of the Sn-rich phase with Bi, the tensile strength tends to be higher than that of the ternary solder. However, while some report similar elongation properties [36,37], ductility has also been observed to decrease together with mechanical fatigue resistance [38]. Bi certainly also improves wettability, which can be significant at lower soldering temperatures or for complex assemblies [39]. Typical compositions are those tested by the Japanese NEDO project Sn–3Bi–2Ag–0.5Cu, and the later IMS project Sn–3Ag–1Bi–0.5Cu. Only fairly low Bi levels are preferred to provide some reduction in melting temperature and improved wetting, while maintaining the desired fatigue life (Bi levels are discussed in Sec. 4.3 on Sn–Ag–Bi solder). Fig. 3.10 illustrates the microstructure of a Sn–Ag–Cu solder with two levels of Bi addition. Particulate Cu–Sn and Ag–Sn intermetallics are formed in the eutectic region in both cases; however, the Sn-rich dendrites formed in the Sn–3Ag–3Bi–0.5Cu samples contain Bi precipitated from the solid solution on aging, making the solder more prone to initial hardening. These precipitates are not typically seen in the alloy containing only 1% Bi. Changes to the microstructure with small Bi additions are not obvious; however, the Bi in higher alloys (e.g., up to 7.5%) is easily seen as this tends to concentrate at the grain boundaries.

Patent considerations will need to be taken into account for Sn–Ag–Cu–Bi solders [1].

3.4.3 Sn–Ag–Bi

Sn–Ag–Bi, with the possible addition of low levels of other elements such as Cu, is a good alternative to Sn–Ag–Cu for products that may demand slightly

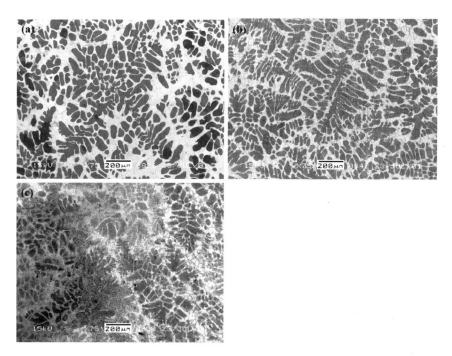

Figure 3.10 Similarity of SEM microstructures of Sn–Ag–Cu–Bi type solders (a) Sn–3Ag–0.5Cu (b) Sn–3Ag–0.5Cu–1Bi, and (c) Sn–3.5Ag–0.5Cu–3Bi. Primary Sn-rich dendrites (the dark phase) together with (lighter) eutectic. Etched with acidic ferric chloride. (From Ref. 38.)

lower production temperatures than are possible with Sn–Ag–Cu, that do not experience high temperatures (e.g., above 90°C) during service, and where there is little possibility of Pb contamination in the joint.

Bi is effective in reducing the solidus temperature of Sn alloys but a little less effective at reducing the liquidus temperature. As a result, Bi containing alloys tend to have a greater pasty range than other lead-free solders, but this remains practically acceptable as long as the Bi content is limited to the recommended range. Experiments have indicated that Sn–3.5Ag alloys with less than 6% Bi added have a single-phase transition with a melting range from 211 to 221°C, while alloys with more than 6% Bi also have a second transition around 137°C [40]. It is recommended that the Bi content is maintained at a low enough level to avoid the possibility of formation of this ternary phase and associated problems of pasty alloys such as fillet lifting. This alloy is in use mainly for the production of consumer goods and a suggested alloy composition range for this solder family is Sn–(3.0–4.0)Ag–(1.0–3.0)Bi, where the melting range can be kept around 210–215°C.

The microstructure consists of Ag_3Sn intermetallic particles, surrounded by a Bi-rich phase, in a Sn-rich matrix. Fig. 3.11 illustrates a typical microstructure of fairly high Bi content, and Fig. 3.12 shows examples of microstructures of Sn–Ag solder joints with a range of Bi levels added; segregated Bi-rich phases become much more evident between 6% and 10% Bi and above [41]. In alloys of limited Bi content, most of the Bi remains in solid solution in the Sn matrix helping to achieve solid solution strengthening immediately after solidification. The high solubility of Bi in Sn at raised temperature is reduced to less than 1% at room temperature. This leads to significant microstructural changes in the aging time following solidification as a result of precipitation of Bi from the Sn-rich phase in order to reach a more stable equilibrium state. This is more evident in high Bi alloys but is always a factor to consider, leading to softening of the solder over time.

As melting temperature is reduced through Bi addition, mechanical properties are also affected, and while tensile strength increases, the ductility of the solder begins to fall off [42]. This is illustrated in Fig. 3.13. This change in mechanical properties has been attributed to the increase of Bi-rich phases with a greater tendency toward brittleness and the solid solution strengthening obtained from the rising Bi content in the Sn-rich matrix. Although ductility and fatigue performance of Sn–Ag–Bi with 4% Bi is

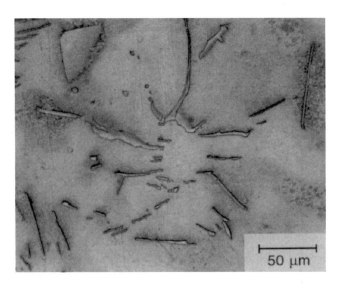

Figure 3.11 Microstructure of Sn–3.33Ag–4.83Bi. (From Ref. 13, with permission from the American Welding Society.)

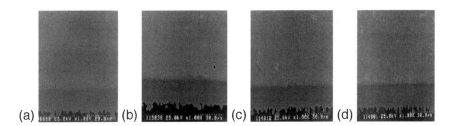

Figure 3.12 Microstructure of Sn–Ag–Bi with variable Bi contents, showing increasing Bi-rich phase (light area) with Bi level Sn–3.5Ag with (a) 3% Bi, (b) 6% Bi, (c) 10% Bi, and (d) 15% Bi. Magnification ×1000. (From Ref. 41, courtesy of T. Baggio, K. Suetsugu [Panasonic], and IPC [Proceedings of IPC Works, Minneapolis, 1999].)

acceptable [43,44] for most applications for which this alloy is recommended for use, greater ductility, and possibly greater fatigue resistance, may be obtained at lower Bi content. At 1% it is reported that Bi remains in solution to provide good properties, but that at higher alloy levels, the formation of clustered precipitates generates stress concentration sites that may reduce fatigue life [45].

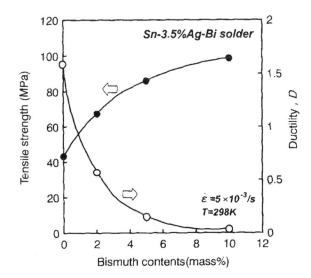

Figure 3.13 Effects of Bi content on tensile strength and ductility of Sn–Ag–Bi. (From Ref. 42.)

Benefits such as improved wetting typically result form the use of low surface tension Bi, enhancing spreading of the solder. However, problems caused by the contamination of Bi alloys with even low levels of Pb can potentially be more serious than those with any other lead-free solder, due to the greater pasty range created (see Table 3.1). It is for this reason that there has been a tendency to avoid Sn–Ag–Bi and similar solders during the transition phase from Sn–Pb to lead-free soldering, but it is likely that the use of Bi will increase when lead-free components and other materials can be guaranteed. The presence of Pb has also been reported to affect and encourage the formation of Bi-rich phases, particularly at grain boundaries, in alloys of slightly higher Bi contents, e.g., 4% Bi, with resultant reduction in performance. Note: high Bi alloys close to eutectic are discussed in Sec. 3.4.6.

The Sn–Ag–Bi solders are subject to some patent restrictions [1].

3.4.4   Sn–Cu–(x)

Sn–Cu is one of the cheaper lead-free alloys available and is often favored in low-cost applications within electronics assembly and for industrial applications such as plumbing. The Sn–Cu binary eutectic forms at around 0.7% Cu and with a eutectic temperature of approximately 227°C. It should be noted that the liquidus rises steeply to high temperatures if the Cu level is increased above 0.7%. For instance, the Sn–3Cu solder sometimes used for industrial applications melts between 227 and 325°C [14]. This can be significant if Cu dissolution is expected from boards in wave soldering operations as the effects on liquidus temperatures can be noticeable and may lead to tight control requirements on the composition of the solder bath.

The microstructure of this alloy consists of a Sn-rich phase with very low levels of Cu, together with Cu_6Sn_5 intermetallic dispersed in a manner dependent on the cooling rate of the solder. Cu_6Sn_5 tends to precipitate in the form of hollow rods, but few of these are seen in practice under conditions of high cooling rate, and small particles are more common. Only limited solid solution strengthening of Sn can be obtained by the addition of Cu because the solution limit is very low indeed, in the region of 0.001%. The microstructure is, in effect, similar to that of the Sn–3.5Ag alloy, but the difference in intermetallic type, size, and dispersion means that the amount of particulate strengthening of the matrix by the particles is lower; this limits the strength of the Sn–Cu eutectic (tensile strength is generally lower than that of Sn–Pb at room temperature) but also contributes to a relatively high ductility over a range of temperatures. In a similar way as in the Sn–Ag–Cu alloy, Cu will be depleted from the matrix close to the area of Cu–Sn intermetallic formation near a Cu substrate interface and will thus limit the growth of the Cu_6Sn_5 intermetallic phase in the bulk of the joint during aging.

The wetting achieved with this solder is sufficient for most purposes, but the relatively high melting point means that more careful choice of flux is required to ensure activation is maintained at appropriate temperatures. Dependent on the wave soldering production conditions, the wetting and flow characteristics of Sn–0.7Cu can be unsatisfactory, with poor penetration of plated-through holes and dull, grainy joints. A particular problem often found with Sn–0.7Cu in wave soldering is bridging, which tends to be resistant to elimination by normal process optimization procedures; attempts to improve fluidity with bath temperatures as high as 300°C have been tried with limited success in bridging reduction, while increasing the probability of board and component damage, and erosion of solder baths, pumps, and nozzles. The addition of Ni and other elements, including low level Ag at 0.5%, has been investigated in order to achieve a greater strength and fatigue resistance of the Sn–0.7Cu eutectic, while at the same time improving the production problems mentioned for the binary alloy. Although Sn–0.7Cu is a eutectic alloy it is likely that primary acicular Cu_6Sn_5 crystals form under nonequilibrium conditions and interfere with solder drainage in the exit area of the wave. The addition of further alloying elements is thought to modify this behavior through incorporation in the crystal structure of the Cu_6Sn_5 phase, with a resulting effect on its surface free energy, and therefore the nucleation and growth behavior of the intermetallic. Because the growth of the intermetallic is a controlling factor during solidification, it could be expected that alloys such as Sn–0.7Cu–0.04Ni would behave differently in the critical exit area of the wave. In practice, levels of around 0.05% Ni in Sn–0.7Cu have been reported to have significant beneficial effects on results obtained during wave soldering. This level is stable and does not require special management during bath replenishment. The alloy is not particularly aggressive toward stainless steel *at the recommended operating temperature*, and no special solder bath materials or coatings have been found to be necessary. However, it is important to ensure that the heating is uniform without surface temperature "hot spots" significantly higher than 260°C [46].

Ni and Cu both have a face center cubic structure and, as a binary mixture, form a continuous solid solution. It is therefore likely that each element is soluble to a reasonable extent within the intermetallic of the other with Sn, i.e., solubility of Ni in Sn–Cu compounds and solubility of Cu in Sn–Ni compounds. While Ni_3Sn_4 and Cu_6Sn_5 have different crystal structures, the Ni_3Sn_2, intermetallic is known to have the same close-packed hexagonal crystal structure as Cu_6Sn_5 with similar lattice constants (as seen in Table 3.3), and as Ni itself has very low solubility in Sn at levels up to about 0.1% virtually all is expected to be incorporated in the Cu_6Sn_5. It has been noted that the addition of Zn to Sn–3.5Ag helps to nucleate fine uniform Ag_3Sn intermetallic precipitates due to high solubility of Zn in Ag compared to Sn;

Table 3.3 Copper and Nickel Intermetallics with Tin

		Cu_6Sn_5	Ni_3Sn_2	Difference
Lattice constants	a	0.4125 nm	0.4190 nm	1.6%
	c	0.5198 nm	0.5086 nm	2.2%
Melting point		415°C	1246°C	831°C

Source: From Ref. 46.

the addition of Ni to Sn–Cu may have a similar effect. The exact intermetallic composition is not known, and, for instance, a variety of phases have been found at interfaces between Ni substrates and Cu containing lead-free solders as shown in Table 2; Ni_3Sn, Ni_3Sn_2, $NiSn_3$ [18,49], and $(Ni,Cu)_3Sn_4$ and $(Ni,Cu)_6Sn_5$ [47,48]. Further investigations are underway.

The Ni addition is also thought to modify the primary Sn phase formation resulting in a finer microstructure as shown in Fig. 3.14, potentially contributing to improved fatigue life. Tensile and elongation properties are generally similar to that of the unmodified solder, but it has been reported that the growth rate of the Cu_6Sn_5 intermetallic is significantly less during aging, providing a more stable structure. Fatigue life has been found to be satisfactory with the Sn–Cu–Ni alloy used in production.

Contamination with Pb is not such a concern as with the Bi containing solders. However, at levels above about 0.5% there is a tendency for a Sn–Pb–Cu eutectic with a melting point of about 182°C to form along grain

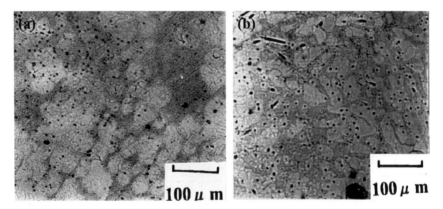

Figure 3.14 Microstructure of (a) Sn–0.7Cu and (b) Sn–0.7Cu–Ni. Cooled from 300°C in a steel mould. Reported refinement of primary dendrite size in the modified alloy; average primary Sn around 50 μm rather than 100 μm. (From Ref. 46, courtesy of Nihon Superior.)

boundaries. This can lead to hot tearing in the fillet as the solder contracts during solidification and has a detrimental effect on long-term thermal fatigue life down to levels around 0.2% [46].

The Sn–0.7Cu binary solder is not subject to patent restrictions but modified alloys such as Sn–0.7Cu–0.04Ni are. Some production licenses have been granted and although the addition of Ni addresses the main deficiency of Sn–0.7Cu, other considerations for best overall performance in wave soldering have led to other trace level additions in proprietary alloys [1].

3.4.5 Sn–Zn–(Bi)

The Sn–Zn binary eutectic forms at around 8.8% Zn and with a eutectic temperature of 198.5°C. These solders have been used in the past in circumstances requiring low melting temperature and corrosion resistance, for example, in brass radiator production, as, in this case, the Zn limits the galvanic corrosion of base metals, e.g., Al in corrosive environments [14]. The use of Zn holds interest for electronic solders due to lower metal cost although many other factors require consideration.

The eutectic microstructure consists of a combination of Sn solid solution with a maximum of 1.7% Zn at the eutectic temperature and a Zn-rich phase containing less than 1% Sn in solid solution [54]. An example is seen in Fig. 3.15. The solubility levels of Zn in Sn are reduced significantly at room

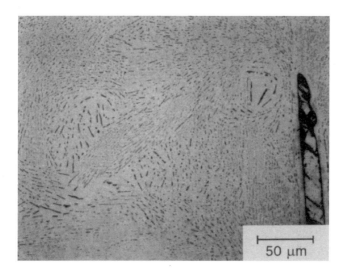

Figure 3.15 Microstructure of Sn–9Zn. (From Ref. 13, with permission from the American Welding Society.)

temperature down to very low (undetermined) percentages. The Sn–Zn binary can be stronger than Sn–Pb at room temperature, but as with all solders this is dependent on solidification rate and microstructure. It is thought that strengthening effects arise from the structure of the Zn-rich phase rather than the solid solution hardening of Zn in the Sn matrix [50–52].

Although Sn and Zn in combination do not form intermetallic compounds, unusually, both are able to and do form intermetallic compounds with Cu. On a Cu substrate a Cu–Zn layer only is often found following soldering, with formation of Cu–Sn following during subsequent aging. In fact, the Cu–Zn system can form three intermediate phases with relatively broad composition ranges that cannot be truly classed as intermetallic compounds; however, both $CuZn$ and Cu_5Zn_8 are typical. The tensile strength of joints can be dramatically reduced after only 100 hr at 150°C as the Cu–Zn layer is broken down and Cu diffuses from the substrate into the bulk of the solder joint to form Cu_6Sn_5 particles [53]. Interdiffusion effects can also lead to voiding at the interface as the rate of diffusion of Zn in Cu is greater than Cu in Zn, encouraging concentration of vacancies at the interface (known as the Kirchendall effect). It is therefore essential to use some form of barrier layer, such as Au/Ni, to prevent these undesirable microstructural changes during aging. Intermetallics of Au–Zn can also form but these are not felt to be as detrimental.

The use of Zn is attractive from an economic point-of-view, being similar in price to Pb, and also due to the effective reduction in melting temperatures of Sn that can be achieved. However, Zn is a relatively reactive element that is practically impossible to use in wave solder systems or other solder baths due to the formation of excessive amounts of very pasty dross— even under inert atmosphere. On the other hand, Zn containing solder pastes can be used for surface mount applications, even while soldering in air, as long as the inevitable limitations on paste shelf-life, printability, and similar factors are deemed to be acceptable. A great deal of research has been carried out in order to provide protection to the solder powder while in the paste to slow reaction times with the flux and other media present and practical systems are in use, but some reduction in printability and stencil life, particularly for fine-pitch, may still be seen with all Zn containing alloys. The wetting achieved by this alloy can be acceptable but is more dependent on successful choice of flux than any other lead-free solder. Specialized methods of solder powder application separate from flux application on a board have been developed to surmount some of the mentioned process difficulties [55].

Additions of Bi are made to the Sn–Zn eutectic in order to lower temperature, improve wetting, corrosion resistance, and other properties. A typical composition is Sn–8Zn–3Bi and an example of a similar microstructure can be seen in Fig. 3.16. Reflow soldering peak temperatures as low as

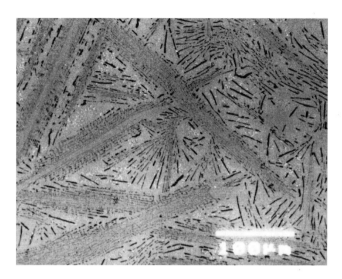

Figure 3.16 Microstructure of Sn–7Bi–6Zn. Bi concentrations can be seen as bright phase. (Courtesy of Soldertec Global at Tin Technology Ltd.)

210°C are possible, with reasonable solderability achieved even in air, and the addition of Bi to the binary alloy appears to improve the practical performance of paste and general corrosion resistance of the solder itself (although not to the general level of other lead-free alloys). While metallic Zn has a high galvanic corrosion potential, this is greatly reduced if the element is not present as a separate pure phase but is contained in solid solution with Sn or even in a Zn-rich phase within an alloy. Currently with Sn–Pb, electrolytic (wet) corrosion is much more common on circuit boards than dry corrosion; however, rate of corrosion also depends on the nobility of the metals involved and their behavior in any electrolyte formed; corrosion of Sn tends to be limited by the formation of a protective passivation layer. Corrosion in the presence of moisture between phases is possible at an exposed surface or at grain boundaries where elemental segregation has occurred or in areas that are cold worked and contain higher energy levels. In both Sn–Zn and Sn–Bi–Zn solders, the Zn phase may be anodic to the remainder of the solder matrix and corrode preferentially if conditions allow, but the surface oxide may provide sufficient protection in these circumstances. This issue has not yet been fully resolved as corrosion behavior is often complex and difficult to predict depending on a variety of environmental conditions such as moisture level, Cl^-, corrosive gases, temperature, oxygen, etc. Work continues on this aspect, and it should be noted that some products such as notebook PCs have been produced with Sn–Bi–Zn solder for several years. Limited tests on Sn–9Zn and Sn–7Bi–6Zn soldered boards exposed to a mixed gas test (30°C, 70% RH, 504 hr) found generation of $ZnCl_2$ deposits on the binary but not the ternary alloy [56,57]. Ionic migration studies of Sn–9Zn and various other lead-free solders have suggested that dissolution of Zn is essentially prevented by the Sn content although not when in contact with flux residues [58].

The binary Sn–9Zn solder is not subject to any patent restrictions but modified alloys will be [1].

3.4.6 Bi–Sn–(Ag)

The Bi–Sn binary system is favored when low soldering temperatures are required, and the alloy holds specific interest for the production of consumer electronics and computing device assembly where maximum service temperatures can be specified.

The eutectic forms at around 57% Bi and with a eutectic temperature of approximately 139°C. The eutectic microstructure consists of solid solution Sn with up to 21% Bi at 139°C and Bi with a very limited level of Sn in solution; intermetallic phases are not formed in the Sn and Bi systems. The high solubility of Bi in Sn is reduced to less than 1% at room temperature, which leads to precipitation of Bi from the Sn-rich phase and associated change in the microstructure and strength during storage. Strengthening of the alloy arises from solid solution hardening effects, and during initial aging, strains across the Sn/Bi phase boundaries as both phases are expanding on aging at room temperature up to 200 hr [59]. On further aging, the precipitation of Bi from the Sn-rich phase and general microstructural coarsening can then lead to alloy softening [40]. In the normal cooling rates experienced in electronics soldering, the microstructure of the alloy will be initially far from the equilibrium situation, and aging effects will be significant as this situation is remedied by Bi precipitation and coarsening.

In some ways the phase diagram and microstructure of Sn–Bi are fairly similar to that of Sn–Pb; however, the mechanical properties are significantly different. Bi itself has several unusual properties that translate into the properties of the eutectic due to the relatively high content of the metal, for instance, the element is brittle, and the eutectic can therefore also suffer from lack of ductility, particularly under conditions of impact and rapid stressing. This is the reason why Sn–Bi is extremely difficult to extrude or draw into wire or strip despite having a high tensile strength when compared to most other lead-free solders. Bi is also unusual in being one of the few metals that expand on solidification. The Sn–Bi eutectic composition itself undergoes little change in volume on solidification, but it does exhibit a slight expansion rather than shrinkage that is found with other alloys, a factor that may create unusual stresses within a solder joint, and which may make removal from a solder pot when solid difficult or impossible.

Bi–Sn is an alloy with high tensile strength but poor ductility when in the cast condition (such as in a solder joint). However, in general, the fatigue performance of Bi–Sn is good, and, notably, Bi–Sn eutectic at temperatures close to its melting point (125°C) has been reported to have the best performance from all lead-free solders tested. It is thought that as the test temperature rises toward the melting point, the diffusion rate increases, which influences creep and enhances crack healing [20]. In a Cu joint, depletion of Sn from the region close to the joint interfacial region in order to form Cu–Sn intermetallic tends to leave the remaining solder region with even greater concentrations of Bi. This can affect the strength of joints making them weaker than may be expected from bulk alloy testing alone. Wetting achieved by these low temperature solders is acceptable, but it should always be remembered that a flux should be selected which is fully active at the soldering temperatures used. Fluxes developed for Sn–Pb alloys activating at 150°C and above are not suitable.

Sn–Bi–Pb forms a ternary eutectic at even lower temperature, 95°C, which can lead to associated process problems (such as fillet lifting), or joint strength reduction (specifically at higher service temperatures, above that of the eutectic formation itself). The composition of this eutectic is around Bi–32Pb–18Sn. The addition of Pb can also affect the microstructure shifting it from eutectic to a combination of primary Sn-rich dendrites, binary Sn/Bi, and the ternary phase [26]. The fatigue life of Bi containing alloys can therefore be reduced through the use of these solders with any other component that may introduce Pb into the joint; for instance, a HASL Sn–Pb-coated circuit board. In the reverse situation, for example, during use of lead-free Sn–Bi component terminations with Sn–Pb solders, only very low levels of Bi are introduced into the joint. This is thought to be acceptable and no reliability problems have been reported due to combination of materials in this way.

Ag is one of a range of alloying additions, which also includes Au, which is considered to be beneficial for fatigue life of Bi–Sn joints. Additions of perhaps 0.5% Ag are made to alloys with an approximate binary eutectic composition and are favored for solders for high specification servers and other computing devices where maximum service temperature can be guaranteed. This small alloying addition produces finer solidification microstructures and specific significant improvements in solder ductility (two to three times that of the binary) and fatigue life [60–62]. The microstructure of

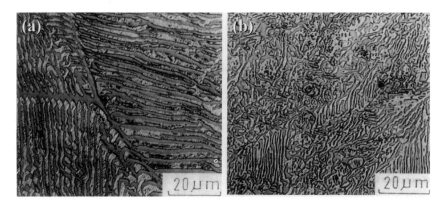

Figure 3.17 Microstructure of (a) Bi–Sn eutectic and (b) Bi–Sn–0.25% Ag showing finer microstructure. (From Ref. 61.)

Bi–Sn and Bi–Sn–Ag can be seen in Fig. 3.17 and improvement in ductility is illustrated in Fig. 3.18. The ductility of the Bi–Sn alloy with low Ag additions is also thought to be less sensitive to strain rate changes than the binary alone, which could indicate superior fatigue performance [6].

The binary Bi–Sn solder is not subject to any patent restrictions but modified alloys will be [1].

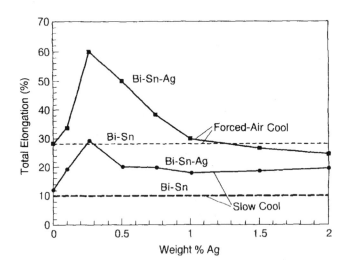

Figure 3.18 Improvement in ductility of Bi–Sn through addition of Ag. (From Ref. 61.)

3.5 OTHER SPECIALIST ALLOYS

3.5.1 Sn–Sb

An Sn–Sb solder may sometimes be used if a slightly higher melting temperature is required, for example, as part of a sequential soldering combination. The alloy is not suggested as an option for typical reflow or wave solder processes. The binary solder is not a eutectic but a peritectic system and therefore cannot form an alloy with a melting temperature lower than Sn. The most commonly used composition is Sn–5Sb with an approximate melting range of 236 to 243°C. This is one of the narrowest melting ranges available from this binary system.

This generally single-phase, solid solution strengthened alloy (seen in Fig. 3.19) is well suited for high-load and relatively high temperature applications and is more typically thought of as a structural solder for industrial use. Any solder with less than 6.7% Sb does not form the large primary Sb_2Sn_3 cubic intermetallics on solidification that can introduce brittleness to solder and joint strength. The high solid solubility of Sb in Sn, over 10% at 240°C, produces significant hardening and strengthening of the solder over the properties of pure Sn or, for instance, Sn–Cu. Some SbSn particles are precipitated during storage at room temperature as the solubility limit is reduced to around 2.5% at that temperature, with associated softening of the

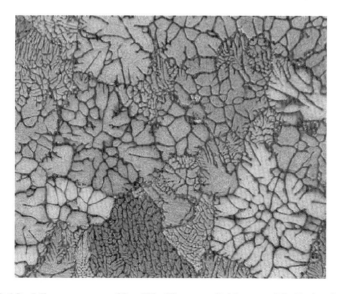

Figure 3.19 Microstructure of Sn–5Sb. (Courtesy Soldertec at Tin Technology Ltd.)

overall alloy [14]. The alloy is particularly resistant to creep. Wetting tends to be somewhat poorer than other lead-free alloys, being more dependent on the use of higher temperatures and correct fluxes.

The binary Sn–Sb binary solders are not subject to any patent restrictions [1].

3.5.2 In–Sn

The Sn–In eutectic of 49% Sn at around 120°C is used in situations where ductility is of particular importance, or where sequential soldering demands a low-temperature solder. In metal itself is very soft, ductile, and has a low tensile strength [63] and imparts some of these properties to the binary system, for example, rapid creep and extensive deformation. The fatigue performance of these solder types is generally poor unless the ductility can be used for a particular advantage. In–Sn is therefore not suggested as a general purpose reflow or wave solder but can be used if scavenging of gold is a problem with other Sn-rich solders that dissolve gold much more rapidly. The use of In bearing solders can also prevent the formation of Sn–Au intermetallics if there is a particular concern over high Au levels from substrates. The Sn–Au intermetallics are brittle and can weaken joints where as the Au–In intermetallics tend to be more tolerant of thermal cycling. Like Sn, In also forms a

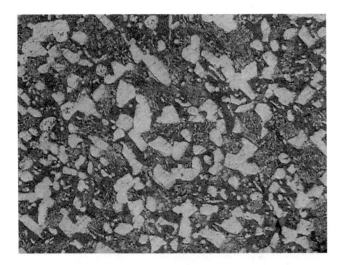

Figure 3.20 Microstructure of In–Sn eutectic solder. Optical micrographs at 250× magnification. The bright phase is tin-rich with a Sn/In ratio of 69:31. The fine darker area is the In-rich phase with a Sn/In ratio of 36:64. (Courtesy of Indium Corporation of America, Research and Development Department.)

good bond with Cu through the formation of Cu–In and Cu–In–Sn inter-metallics, for example, Cu_2In_3Sn [4]; however, dissolution of Cu can be so rapid that a barrier layer of Ni often has to be used [64,65].

The eutectic microstructure consists of Sn- and In-rich phases each with high solid solubility and can be seen in Fig. 3.20. Data quoted vary, but the phase diagram suggests 56–88% In in the In-rich phase and 71–95.7% Sn in the Sn-rich phase. Little or no change in microstructure through, e.g., coarsening has been observed in this alloy during deformation or fatigue. This is probably due to the recovery that must occur during deformation at such temperatures close to the melting point [65].

In-based solders have several other notable properties. For instance, it is possible to wet ceramic and glass without the need for solderable finishes; In–Sn also maintains reasonable ductility at very low temperatures such as $-150\,^\circ$C and may be a possible alloy for use in cryogenic applications [66].

The binary Sn–In binary solders are not subject to any patent restrictions [1].

3.6 COSTS

All lead-free alloys have a higher metal cost per ton because the Pb that is being replaced is one of the cheapest metals available. However, there is a significant difference between the density of Sn–Pb (8.4 g/cm^3) and that of high Sn alloys (7.3 g/cm^3), which in some way compensates for this increased cost. The most appropriate comparison of metal costs must be calculated on the basis of equivalent volume rather than weight. Metal cost per cubic centimeter is provided in Table 3.4 where the effects of additions of high value metals such as Ag and In can be seen. It should also be noted that the price of some metals such as In is more subject to speculation and significant price fluctuations. The costs are not current but an indication only.

It is important to note, however, that the final cost of a solder product is not just dependent on metal cost but is the result of an accumulation of additional factors that have an impact on the final product charge. For example, although Sn–Zn-based solders may have a lower metal base cost, the additional technological development of measures required to protect these reactive solder powders and achieve satisfactory shelf-life will add to product price.

Patent fees, and the resulting license fees payable to the patent holder, can also influence solder costs. While many agreements between solder-producing companies have been achieved satisfactorily at nominal costs, certain patents on Sn–Ag–Cu have been a major stumbling block in the development of lead-free technology. This has had a significant effect on the

Table 3.4 Metal Cost of Alloys Relative to Sn–37Pb

	U.S. cents per cm^3	Multiple of Sn–37Pb cost
Sn–3.5Ag	10.12	2.05
Sn–3.4Ag–0.8Cu	9.98	2.02
Sn–3Ag–1Bi–0.5Cu	9.57	1.94
Sn–3Ag–3Bi	9.61	1.95
Sn–0.7Cu	6.27	1.28
Sn–9Zn	5.74	1.16
Sn–8Zn–3Bi	5.82	1.18
Bi–43Sn	6.66	1.35
Bi–42Sn–0.5Ag	7.30	1.48
Sn–5Sb	6.07	1.23
In–48Sn	76.91	15.60
Pb–37Sn	4.93	–

Does not include development, license or manufacturing costs. Comparison of cost is by volume not weight as this is more appropriate to solder joint size. Metal costs are variable and provide an approximate comparison only.

choice of exact alloy composition within a solder family and will also directly affect solder price.

3.7 SELECTION RECOMMENDATIONS

Examples of the type of considerations to be made during alloy selection procedures for a wave soldering process and a reflow soldering process are illustrated in Figs. 3.21 and 3.22. The wave solder selection diagram begins with the Sn–Cu solder for reasons of cost and, through a series of questions, outlines why other choices may be more preferable. The reflow solder selection diagram begins at the Sn–Ag–Cu ternary solder, as basic metal cost is less of an issue with paste supply, and indicates some lower temperature alloys that may be used if required. These are very approximate guidelines that can be used to assist in the selection of an alloy, but they do not incorporate all possible influencing factors, which should be examined in association with knowledge of the intended application and details such as joint design, substrate type, processing conditions, etc. These guidelines should only be used in combination with comments provided in the sections on individual alloy types and with further detailed research, to ensure that the appropriate application specific alloy choice can be made.

Various recommendations have been made by organizations within the electronics industry since the first research program on lead-free electronic

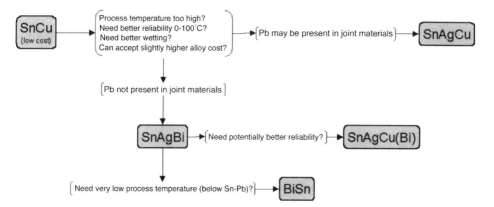

Figure 3.21 Example flow chart for selection of a wave soldering alloy. (Adapted from an idea of G Reichelt, Technolab. From Ref. 77, courtesy of Soldertec Global at Tin Technology Ltd.)

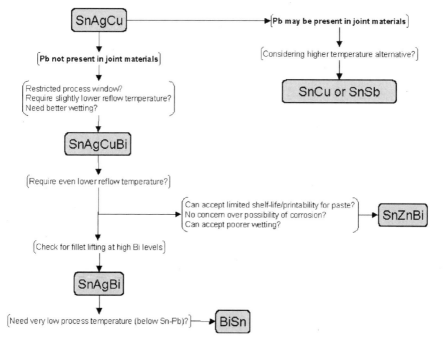

Figure 3.22 Example flow chart for selection of a reflow soldering alloy. (Adapted from an idea of G Reichelt, Technolab. From Ref. 77, courtesy of Soldertec Global at Tin Technology Ltd.)

solders were completed in the early 1990s. One of the first was a U.K. Government (DTI)-sponsored project that ran from 1991 to 1994 involving the then named companies GEC-Marconi, BNR Europe, ITRI, and Multicore, which made initial investigations on a range of alloys including Sn–Ag–Cu–Bi, Sn–Ag–Bi, Sn–Zn–Bi, and others [56,57]. This was used as the basis of the European-funded program on lead-free from 1996 to 1999, IDEALS (Improved Design life and Environmentally aware manufacture of Electronic Assemblies by Lead-free Soldering), involving companies then named GEC-Marconi, Philips, Siemens, Witmetaal, Multicore, and NMRC. This consortium completed investigations of both reflow and wave soldering with lead-free and concluded that Sn–3.8Ag–0.7Cu was the optimum lead-free alloy "functionally equivalent" in performance to Sn–Pb. The solders Sn–Ag–Bi and Sn–Ag–Cu–Sb were also felt to be suitable for use in single-sided wave soldering and all wave soldering processes, respectively [67].

In the United States, the NCMS (National Center for Manufacturing Sciences) lead-free program was initiated in 1993 and involved 11 corporations, academic institutions, and national laboratories. On completion in 1997, after examination of perhaps 70 solders, the consortium made a series of application-specific alloy choice recommendations. These are summarized in Table 3.5, taken from the final project report. It can be seen that Bi–42Sn, Sn–3.5Ag–4.8Bi, and Sn–3.5Ag were recommended. Sn–Ag–Cu alloys were not examined in detail [20].

In October 1999, Soldertec (the solder focused division of Tin Technology) recommended that Sn–Ag–Cu should be considered as the mainstream lead-free solder choice for all process types, reflow, wave, and hand soldering. This was concluded from examination of available data and in consultation with leading solder manufacturing companies. No specific alloy was recommended but a composition range of Sn–[3.4–4.1]Ag–[0.45–0.9]Cu. This range was due in part to consideration of patent restrictions and allowed the inclusion of most of the important compositions used at the time. This was also thought to be the approximate region of the phase diagram where roughly equivalent properties could be achieved. Other alloys were also mentioned as alternative choices depending on circumstances, for example, a lower temperature alloy or one of lower cost may be preferred. The Soldertec recommendation of Sn–Ag–Cu did not preclude the use of other alloys such as Sn–Cu, Sn–Ag–Bi, Sn–Ag, and Sn–Zn–Bi, and it was anticipated then, as it is now, that users able to make informed choices on the differences between these alloys would prefer to use one or more for specific purposes, particularly when Pb had been totally eliminated from the process [68].

In the United States, the NEMI consortium of around 50 companies, suppliers, government agencies, and universities (National Electronics Manufacturing Initiative) followed this in January 2000 with a recommendation of a specific alloy, Sn–3.9Ag–0.6Cu, as the primary choice for reflow soldering.

Table 3.5 Alloy Selection from National Centre for Manufacturing Sciences Project, U.S.A., 1997

Alloys	Liquidus and solidus temperatures	Industry sectors	Evidence for recommendation
Bi–42Sn	139°C (eutectic)	Consumer electronics	Simple two-component eutectic alloy but low eutectic temperature restricts maximum use temperature.
		Telecommunications	*Surface mount technology:* Better fatigue life than eutectic tin–lead for both thermal cycling ranges; less fatigue damage than eutectic tin–lead in surface mount cross-sections.
			Through-hole technology: Mixed results with fatigue life for CPGA-84 better than eutectic tin–lead; for CDIP-20 worse than eutectic tin–lead.
Sn–3.5Ag–4.8Bi	210 to 205°C	Consumer electronics Telecommunications Aerospace Automotive	*Surface mount technology:* Longer fatigue life than eutectic tin–lead at 0 to 100°C; no failures in 1206 resistors up to 6673 cycles; fatigue life equivalent to eutectic tin–lead at −55 to +125°C; less fatigue damage than eutectic tin–lead seen in surface mount cross sections.
			Through-hole technology: Most joints show failure by fillet lifting.
Sn–3.5Ag	221°C (eutectic)	Consumer electronics Telecommunications Aerospace Automotive	Simple two-component eutectic alloy. *Surface mount technology:* Fatigue life equivalent to eutectic tin–lead at 0 to 100°C; worse than eutectic tin–lead at −55 to 125°C.
			Through-hole technology: Less susceptible than other high-tin solders to fillet lifting but results from Sn–2.6Ag–0.8Cu–0.5Sb indicate through-hole reliability still may be compromised.

Source: Ref. 20.

Table 3.6 Alloy Selection from JEIDA Roadmap on Lead-Free Soldering, Japan, 2000

Process		Alloys	Recommended composition	Remarks
Wave		Sn–3.5Ag Sn–(2–4)Ag–(0.5–1)Cu Sn–0.7Cu + additives (Ag, Au, Ni, Ge, In)	Sn–3Ag–0.5Cu	SnPb plating metals on components might cause fillet-lifting and damage to boards.
Reflow	Medium and high temperature	Sn–3.5Ag Sn–(2–4)Ag–(0.5–1)Cu Sn–(2–4)Ag–(1–6)Bi including with 1–2% In Sn–(8–9)Zn–(0–3)Bi	Sn–3.0Ag–0.5Cu Sn–8Zn–3Bi	Needs temperature control as SnAg melts at high temperature. Incompatible with SnPb plated components when containing Bi. Handle carefully SnZn in corrosive environment. NiAu finishes preferred for Cu electrode at high temperature.
	Low temperature	Sn–(57–58)Bi	Sn–57Bi–1Ag	Incompatible with SnPb plated components.
Manual/robot (thread solder)		Sn–3.5Ag Sn–(2–4)Ag–(0.5–1Cu) Sn–0.7Cu + additive (Ag, Au, Ni, Ge, In)	Sn–3Ag–0.5Cu	Incompatible with different solder alloys in reworking.

Source: Ref. 70.

However, this alloy was not selected for wave soldering in which Sn–0.7Cu was favored, with Sn–3.5Ag as a secondary alternative [69].

A significant amount of work has, of course, also been carried out in Japan from the 1990s onward regarding selection of suitable lead-free solder compositions. The JEIDA Lead-Free Roadmap also published in the year 2000 recommended the same alloy family but with a slightly lower Ag content, Sn–3.0Ag–0.5Cu, together with two other notable solder alloys, the Sn–Zn–Bi for low temperature use and Bi–Sn–Ag for very low temperature use. The JEITA alloy selections can be seen in Table 3.6 [70].

In May 2002, the IPC solder Value Council, consisting of major global solder alloy manufacturers, pin-pointed Sn–3.0Ag–0.5Cu as a baseline from which further work should progress in order to finalize a favored Sn–Ag–Cu composition. The individual solder manufacturers currently remain free to, and do, recommend alternative versions. This low Ag composition is not necessarily considered as the final selection but as a starting point for discussion [71].

It was felt by each of these organizations when providing this type of information that the industry would benefit from being able to focus on one alloy type for research, development, and implementation purposes, and that such recommendations would be able to achieve this goal. However, up until the time of writing, alloy selections in the different regions of the world have yet to converge and no globally agreed recommendation exists. In part, this is due to the evolution of factors as the transition to lead-free is made.

3.8 COMPOSITIONAL TOLERANCES

All solder compositions are permitted a certain tolerable variation within which properties are not expected to vary significantly. The J-STD-006 (IPC, May 2001) [72] allows a tolerance of $\pm 0.20\%$ of the nominal percentage if that is less than 5.0%, or $\pm 0.50\%$ when the nominal percentage is greater than 5.0%. For example, Sn–3.5Ag has a nominal percentage of 3.5% with an allowable tolerance of $\pm 0.20\%$. The allowable compositional range is therefore 3.3% to 3.7%. The current version of ISO standard 9453 (1990) [73] specifies compositional ranges for each alloy type. For example, Sn–4Ag is permitted to have an Ag content in the range 3.5–4.0%. This document is currently under review.

3.9 PB CONTENT OF "LEAD-FREE" ALLOYS

The J-STD-006 (IPC, May 2001) [72] has definitions for both 0.2% and 0.1% Pb in solder alloys according to specific customer designation. For example, variation E designation demands a Pb content below 0.10% and an Sb content below 0.20%. The current version of ISO standard 9453 (1990) [73] specifies Pb content limits for each lead-free alloy. For example, Sn–5Sb, Sn–1Cu, Sn–4Ag-type solders are permitted to contain up to 0.10% Pb. However, Bi–43Sn and Sn–50In are permitted to contain only 0.05% Pb. This document is currently under review. It should be remembered that the Pb impurity content may be more important for certain lead-free alloys (Bi containing) than others to ensure compatibility problems are minimized.

There is some indication of Pb impurity level definition within EU legislation and potential levels to be expected in the RoHS Directive [74] since through examination of a Commission Decision [76] regarding a Pb ban in End of Life Vehicles [75], it can be seen that 0.1 wt.% Pb is a threshold level for lead-free in this application. "A concentration of up to 0,1% by weight and per homogeneous material, of Pb. . .shall be tolerated, provided these substances are not intentionally introduced. The above thresholds are in line with EC legislation on hazardous substances and preparations and also take into account the need not to hamper the recycling of waste (such as secondary aluminium)." Intentional addition is defined as the deliberate use of an element "in the formulation of a material or component where its continued presence is desired in the final product to provide a specific characteristic, appearance or quality." Contamination from recycled material is not classed as intentional introduction but must be below the specified 0.1% limit.

It is logical to assume that a similar threshold level will be proposed in the RoHS Directive [74] in due course, unless specific technical differences can be demonstrated between the case of Pb in vehicles (including electronics) and Pb in products covered by the RoHS proposals.

ACKNOWLEDGMENTS AND DISCLAIMER

The author would like to thank all those who assisted with the preparation of this chapter and the support of the ITRI tin producing member companies who have supported the development of lead-free technology over many years.

We believe the information provided in this statement and any attachments is reliable and useful, but it is furnished without warranty of any kind from the authors. Potential users should make their own determination of the suitability of any information provided and adopt any safety, health, and other precautions as may be deemed necessary by the user. No license under any patent or other propriety rights is granted or to be inferred from the provision of the information herein. In no event will Tin Technology or any of its affiliates be liable for any damages whatsoever resulting from the use of or reliance upon this information.

REFERENCES

1. Note that specific legal advice on patent cover for any solder alloy in any region should be taken. Previously known and used binary Sn alloys are normally not affected by patent issues.
2. M Abtew, G Selvaduray. Lead-free solders in microelectronics. Mater Sci Eng 27:95–141, 2000.
3. J Glazer. Microstructure and mechanical properties of lead-free solder alloys for low-cost electronic assembly: A review. J Electron Mater 23(8):670–693, 1994.
4. J Glazer. Metallurgy of low temperature lead-free solders for electronic assembly. Int Mater Rev 40(2):65–93, 1995.
5. T Siewert, S Liu, S Smith, JC Madeni. Database for Solder Properties with Emphasis on New Lead-Free Solders. Colorado: NIST and Colorado School of Mines, February 2002.
6. EP Wood, KL Nimmo. In search of new lead-free electronic solders. J Electron Mater 23(8):709–713, 1994.
7. KL Nimmo. Review of current issues in lead-free soldering. Proceedings of Surface Mount International, San Jose, 7–11 September 1997.
8. SK Kang, AK Sarkhel. Lead (Pb)-free solders for electronic packaging. J Electron Mater 23(8):701–707, 1994.
9. AZ Miric, A Grusd. Lead-free alloys. Solder Surf Mt Technol 10(1):19–25, 1998.
10. NC Lee. Getting ready for lead-free solders. Solder Surf Mt Technol 9(26):65–69, 1997.
11. K Suganuma. Research projects and promotion of lead-free soldering in Japan. Workshop at EcoDesign 2001, Tokyo, 15 December 2001.
12. Solder Alloy Data: Mechanical Properties of Solders and Solder Joints, Publication Number 656. London: International Tin Research Institute, 1986.
13. PT Vianco. Soldering Handbook. 3rd ed. Miami: American Welding Society, 1999, pp 207–228
14. The Soft Soldering Handbook. 5th ed. London: Tin Technology, 2001, pp 19–31.
15. JS Hwang. Environment-friendly electronics: Lead-free technology. Isle of Man: Electrochemical Publications, 2001.
16. J Haimovich. Intermetallic compound growth in tin and tin–lead platings over nickel and its effects on solderability. Weld Res Suppl 102-s–111-s, 1989.
17. MS Lee, CM Liu, CR Kao. Interfacial reactions between Ni substrate and the component Bi in solders. J Electron Mater 28(1):57–62, 1999.
18. SK Kang, RS Rai, S Purushthaman. Interfacial reactions during soldering with lead–tin eutectic and lead-free tin-rich solders. J Electron Mater 25(7):1113–1120, 1996.

19. KL Lin, CJ Chen. The interactions between In–Sn solders and electroless Ni–P deposit upon heat treatment. J Mater Sci 7:397–401, 1996.
20. National Centre for Manufacturing Sciences NCMS Lead-Free Solder Final Report 0401RE96. Ann Arbor: NCMS, August 1997.
21. D Shangguan, A Achari. Lead-free solder development for automotive electronics packaging applications. Proceedings of SMI, San Jose, 1995, pp 423–428.
22. C Melton. Nitrogen atmosphere processing in lead-free soldering. Proceedings of Nepcon West, USA, 1995, pp 1003–1011.
23. HD Solomon. Low cycle fatigue of Sn96 solder with reference to eutectic solder and a high PB solder. J Electron Packag 113:102–108, 1991.
24. S Chada, W Laub, RA Fournelle, D Shangguan, A Achari. Microstructural evolution of Sn–Ag solder joints resulting from substrate copper dissolution. Proceedings of the SMTA Conference, San Jose, September 1999.
25. S Chada, A Hermann, W Laub, R Fournelle, D Shangguan, A Achari. Microstructural investigation of Sn–Ag and Sn–Pb–Ag solder joints. Solder Surf Mt Technol 26:9–13, 1997.
26. A Gickler, C Willi, M Loomans. Contamination of lead-free solders. SMT, 44–48, November 1997.
27. R Gehman. Dendritic growth evaluation of solder thick films. Int J Hybrid Microelectron, 6:239–242.
28. CD Miller, IE Anderson, JF Smith. A viable Sn–Pb solder substitute: Sn–Ag–Cu. J Electron Mater 23(7):595–601, 1994.
29. WQ Peng. Lead-free Electronic Assembly Based on Sn–Ag–Cu Solders. Licentiate thesis, Helsinki University of Technology, Helsinki, 2001.
30. ME Loomans, ME Fine. Tin–silver–copper eutectic temperature and composition. Metall Mater Trans A 31(4):1155–1162, 2000.
31. KW Moon, WJ Boettinger, UR Kattner, FS Biancaniello, CA Handwerker. Experimental and thermodynamic assessment of Sn–Ag–Cu solder alloys. J Electron Mater 29(10):1122–1136, 2000.
32. KS Kim, SH Huh, K Suganuma. Effects of cooling speed on microstructure and tensile properties of Sn–Ag–Cu alloys. Mater Sci Eng, A Struct Mater: Prop Microstruct Process A333(1–1):106–114, 2002.
33. Unpublished data. Tin Technology Ltd.
34. K Zheng, V Vuorinen, JK Kivilahti. Some reliability issues of lead–free soldering. Seminar of Helsinki University of Technology, November 2000.
35. E Gebhardt, G Petwoz. Constitution of the Ag–Cu–Sn system. Z Metallkde 50(10):597–605, 1959.
36. M Okamoto, H Shimokawa, K Serzawa, T Narita. Reliability of joints between lead-free solder and lead-free metallized leads. Tokyo: EcoDesign, 2001, pp S193–S198.
37. J Hwang. A strong lead-free candidate: the Sn/Ag/Cu/Bi system. Surf Mt Technol, August 2000.
38. C Kanchanomai, Y Miyashita, Y Mutoh. Low cycle fatigue behavior of Sn–Ag, Sn–Ag–Cu, and Sn–Ag–Cu–Bi lead-free solders. J Electron Mater 31(5):456–465, May 2002.

39. M Miyazaki, S Nomura, T Takei, N Katayama, H Tanaka, M Akanuma. Upgrading lead-free soldering technology (2). Tokyo: EcoDesign, 2001, pp S187–S192.

40. F Hua, J Glazer. Lead-free solders for electronic assembly. Proceedings of Design and Reliability of Solders and Solder Interconnections, Orlando, 1997, pp 65–73.

41. T Baggio, K Suetsugu. The Panasonic Mini-Disk Player: Turning a new leaf in a lead-free market. Proceedings of IPC Works, Minneapolis, 1999, pp S-05-4-1–S-05-4-8.

42. Y Kariya, M Otsuka. Effect of bismuth on the isothermal fatigue properties of Sn–3.5 mass% Ag solder alloy. J Electron Mater 27(7):866–870, 1998.

43. PT Vianco, JA Regent. Tin–silver–bismuth solder for electronics assembly. US Patent 5,439,639. August 1995.

44. P Vianco, JA Regent, I Artaki, U Ray. An evaluation of prototype circuit boards assembled with a Sn–Ag–Bi solder. Proceedings of IPC Works, Minneapolis, 1999, pp S-03-3-1–S-03-3-34.

45. Z Guo, TJ Gher. The strengthening and dragging characteristics of dilute Sn solders designed for electronic packages. Manuf Sci Eng 4:176–198, 1996.

46. Personal communication, Nihon Superior, Japan, 2002.

47. V Vuorinen, JK Kivilahti. On the formation of the intermetallic compounds in the Sn–Cu–Ni system. Proceedings of Postgraduate Seminar: Solid–Solid and Gas–Solid Reactions, Helsinki, 1997, pp IIX/1–IIX/5.

48. PG Harris, KS Chaggar. The role of intermetallic compounds in lead-free soldering. Solder Surf Mt Technol 10(3):38–52, 1998.

49. SK Kang, V Ramachandran. Scr Metall 14:421–424, 1980.

50. M Tadauchi, I Komatsu, H Tateishi, K Teshima. Sn–Zn eutectic alloy soldering in a low oxygen atmosphere. Tokyo: EcoDesign, 2001, pp 1055–1058.

51. F Vnuk, M Sahoo, D Baragar, RW Smith. Mechanical properties of the Sn–Zn eutectic alloys. J Mater Sci 15:2573–2583, 1980.

52. M Suzuki, H Matsuoka, E Kono. Applications of Sn–Zn lead-free solder in reflow soldering process. Tokyo: EcoDesign, 2001, pp S175–180.

53. K Suganuma, T Murata, H Noguchi, Y Toyoda. Heat resistance of Sn–9Zn solder/Cu interface with or without coating. J Mater Res 15(4):884–891, April 2000.

54. M McCormack, S Jin. Improved mechanical properties in new lead-free solder alloys. J Electron Mater 23(8):715–720, 1994.

55. T Shoji, Y Kinoshita, T Kuramoto. Highly reliable lead-free solder pastes of Sn–Zn alloy system. Proceedings of IPC Works, Minneapolis, 1999, pp S-03-7-1–S-03-7-7.

56. PG Harris, MA Whitmore. Alternative solders for electronics assemblies. Part 1. Circuit World 19(2):25–27, 1993.

57. PG Harris, MA Whitmore, JH Vincent, BP Richards, DR Wallis, IA Gunter, M Warwick, HAH Steen, SR Billington, AC Harman. Alternative solders for electronics assemblies. Part 2. Circuit World 19(3):32–34, 1993.

58. S Yoshihara, H Tanaka, F Ueta, K Kumekkawa, H Hiramatsu, T Shirakashi.

Newly developed real-time monitoring system for ionic migration of lead-free solder by means of quartz crystal microbalance. Tokyo: EcoDesign, 2001, pp 1064–1068.

59. EW Hare, R Corwin, EK Riemer. Structure and property changes during room temperature aging of bismuth solders. Electron Packag Mater Process, 1986, pp 109–115.

60. F Hua, Z Mei, J Glazer, A Lavagnino. Eutectic Sn–Bi as an alternative lead-free solder. Proceedings of IPC Works, Minneapolis, 1999, pp S-03-8-1–S-03-8-6.

61. M McCormack, HS Chen, GW Kammlott, S Jin. Significantly improved mechanical properties of Bi–Sn solder alloys. J Electron Mater 26(8):954–958, 1997.

62. Y Yamagishi, H Ueda, M Kitajima. Lead-free solder. Fujitsu 48(4):305–309, 1997.

63. M Plotner, B Donat, A Benke. Deformation properties of indium-based solders at 294 and 77K. Cryogenics 31:159–162, 1991.

64. JL Freer Goldstein, JW Morris. The effect of substrate on the microstructure and mechanical behavior of eutectic tin–indium. Proceedings of the Materials Research Society Symposium 323, 1994, pp 159–164.

65. JL Freer Goldstein, JW Morris. Microstructural development of eutectic Bi–Sn and eutectic In–Sn during high temperature deformation. J Electron Mater 23(5):477–486, 1994.

66. W Kinzy Jones, Y Liu, M Shah, R Clarke. Mechanical properties of Pb/Sn Pb/ In and Sn–In solders. Solder Surf Mt Technol 10(1):37–41, 1998.

67. IDEALS—Improved design life and environmentally aware manufacturing of electronics assemblies by lead-free soldering. Brite-Euram Contract BRPR-CT96-0140. Project number BE95-1994. 1 May 1996 to 30 April 1999.

68. Lead-free alloys—the way forward. London: Soldertec, October 1999, pp 1–2.

69. NEMI Group recommends tin/silver/copper alloy as industry standard for lead-free solder reflow in board assemblies. Herndon, VA: NEMI Press Release, January 2000, pp 1–2.

70. JEIDA Roadmap on Lead-Free Soldering. Tokyo: JEIDA, 2000.

71. K Seelig. Seminar at Lead-Free Electronic Components and Assemblies, San Jose, 1–2 May 2002.

72. IPC/EIA J-STD-006A. Northbrook, Il: IPC, May 2001.

73. ISO 9453. International Standard, ISO, 1990-12-01.

74. EU Directives on Waste from Electrical and Electronic Equipment (WEEE), 2002/96/EC, 13 February 2003, and, Restriction of Hazardous Substances in Electrical and Electronic Equipment (RoHS), 2002/95/EC, 13 February 2003.

75. EU Directive, End of Life Vehicles, 2000/53/EC, 18 September 2000.

76. EU Commission Decision, Amending Annex II of 2000/53/EC End of Life Vehicles Directive, 2002/525/EC, 27 June 2002.

77. Soldertec website, http://www.lead-free.org.

4

Plating Lead-Free Soldering in Electronics

Hidemi Nawafune

Konan University, Kobe, Japan

Lead-free soldering is an urgent matter in the electronics industry. In order to achieve lead-free soldering in electrical appliances, securing the bonding reliability in mounting electronic parts is indispensable, and the surface treatment of the electrode in the electronic parts is a major technology in addition to the development of lead-free solder. From such a viewpoint, studies concerning lead-free plating of Sn or Sn alloy such as Sn–Ag, Sn–Bi, Sn–Cu, Sn–Zn, etc., which include Sn as a matrix, on the electrode of electronic parts were aggressively carried out to impart solderability, and introductions and explanations have been reported recently [1–5]. However, neither lead-free soldering nor a plating process corresponding to soldering has been adopted by each electric manufacturer under a unified concept at present. In this report, the basic science of Sn and Sn-based alloy plating, the features of each plating bath, and film properties will be introduced.

4.1 BASIS OF PLATING

The composition and the crystal structure of the plating film are closely related to the deposition mechanism. In particular, the deposition reaction of alloy plating is affected by many factors. However, it is too complicated to consider all of these factors when considering the deposition mechanism of alloy plating. Usually, the following method is applied: The deposition process of an individual alloy component is examined, that is, the equilibrium potential in the deposition of an individual alloy component is compared, and the deposition mechanism of the alloy is considered. When the equilibrium potentials are close, as in the case of the Sn–Pb alloy, an alloy film can be obtained comparatively easily.

4.1.1 Metal and Alloy Plating from a Simple Salt Bath

The equilibrium potential of the deposition reaction of the metal can be obtained by the Nernst equation based on the electrochemical equilibrium. Equation (4.1) expresses the equilibrium potential E [V] in the deposition reaction ($M^{n+} + ne^- \rightarrow M$) of the metal from a simple salt bath, that is, from the metal ion under the condition of an aqua complex (ligand:water). At $25°C$, the relation is expressed by Eq. (4.2).

$$E = E^0 + (RT/nF)\ln(aM^{n+}/aM) \tag{4.1}$$

$$E = E^0 + (0.059/n)\log(aM^{n+}/aM) \tag{4.2}$$

Here, E^0 is the standard electrode potential [V], R is the gas constant [8.3143 $C \cdot V/mol \cdot K$], T is the absolute temperature [K], n is the number of electrons involved in the discharge of the ion, F is Faraday's constant [96485 C/mol], aM^{n+} is the activity of M^{n+}, and aM is the activity of M. M can be considered equal to 1 for a monometal or a eutectic alloy such as lead-free solder.

In Sn–Pb alloy plating from a simple salt bath such as the current methanesulfonate bath, the standard electrode potential of tin ($Sn^{2+} + 2e^- \rightarrow Sn$) is -0.138 V and that of Pb ($Pb^{2+} + 2e^- \rightarrow Pb$) is -0.129 V, and the difference in the standard electrode potential is only 0.009 V. In such a case, alloy film of an arbitrary composition can be obtained by adjusting the concentration of the Sn^{2+} and Pb^{2+} ions in the plating bath. However, for Sn–Cu alloy plating from a simple salt bath, the standard electrode potential of Cu ($Cu^{2+} + 2e^- \rightarrow Cu$) is 0.337 V and is extremely noble compared to that of Sn. From Eq. (2), the equilibrium potential of Sn is $E = -0.138 + 0.03\log a_{Sn^{2+}}$, and the equilibrium potential of Cu is $E = 0.337 + 0.03\log a_{Cu^{2+}}$; therefore, Cu seems to be extremely easily deposited, and the alloy deposition does not seem to take place. If their equilibrium potentials are adjusted to be equal by adjusting the concentration (i.e., activity, to be exact), it is necessary to make the concentration of Cu^{2+} ion $1/10^{16}$ of the concentration of Sn^{2+} ion. Consequently, the deposition of the Sn–Cu alloy from the bath with this Cu^{2+} ion concentration is impossible.

As clearly understood from the standard electrode potential of the metals, which are examined for practical lead-free solder plating, as shown in

Fig. 4.1, the alloy deposition is difficult in Sn–Ag alloy plating for the same reason. On the contrary, in Sn–Zn alloy plating, it is expected that the preceding deposition of Sn, which has a nobler standard electrode potential, will occur, and the alloy deposition is also difficult. Moreover, it is well known that the preceding deposition of Bi, which has a nobler standard electrode potential, occurs in Sn–Bi alloy plating where their standard electrode potentials are similar, and Bi is deposited by substitution on the anode and the cathode (parts to be plated).

4.1.2 Alloy Plating from a Complex Bath

Conventionally, a complex bath has been used for the deposition of an alloy. Equation (4.3) expresses the equilibrium potential in the deposition reaction $(ML^{n-mx} + ne^- \rightarrow M + xL^{m-})$ of the metal from the complex bath, that is, from the metal complex ion. It can be formulated as Eq. (4.4) at 25°C.

$$E' = E^{0\prime} + (RT/nF) \ln aMLx^{n-mx}/[(aM^{n+}/aL^{m-})^x] \tag{4.3}$$

$$E' = E^{0\prime} + (0.059/n) \log aML^{xn-mx}/[(aM^{n+}/aL^{m-})^x] \tag{4.4}$$

Here, $E^{0\prime}$ is the standard electrode potential $[E^{0\prime} = E^0 - (RT/nF) \ln K]$ of the electrode reaction of the complex ion, and K is the stability constant $(K = [ML^{xn-nx}]/([M^{n+}][L^{m-}]^x))$ of the metal complex ion ML^{xn-mx}. Although Eqs. (4.3) and (4.4) express the equilibrium potential of the deposition reaction of the metal from the metal complex ion ML^{xn-mx}, the metal complex ion in the complex bath is not one kind but a mixed system of multiple ion species. Therefore, it is necessary to consider the ion species that most closely relates to the deposition, not the complex ion that is the main component in the bath, to obtain the equilibrium potential. The concentration of the ion species largely changes with the pH and the concentration of the complexing agent. Moreover, because the ion species that directly participates in the deposition might change to another ion species due to the change in the

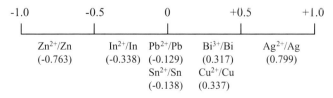

Figure 4.1 Standard electrode potential of metals related to lead-free solder plating.

pH and the bath composition, it is not easy to obtain the equilibrium potential of the deposition reaction in the complex bath accurately.

For instance, up to two pyrophosphoric acid ions can be coordinated to a Cu^{2+} ion in the copper–pyrophosphate complex. The copper complex ion in the bath is mainly two pyrophosphoric acid coordinated $[Cu(P_2O_7)_2]^{6-}$ ions, but the species that directly participates in the deposition is one pyrophosphoric acid coordinated $[CuP_2O_7]^{2-}$ ion. Consequently, the equilibrium potential of Cu deposition from the pyrophosphate bath ($CuP_2O_7^{2-} + 2e^- \rightarrow Cu + P_2O_7^{4-}$) is obtained from the following equation.

$$E' = E^{0'} + 0.03 \log aCuP_2O_7^{2-}/(aCu \cdot P_2O_7^{4-}) \qquad (4.5)$$

The standard electrode potential $E^{0'}$ of the electrode reaction of the complex ion can be obtained from $E^{0'} = E^0 - (RT/nF) \ln K$. Substituting $E^0 = 0.337$ V, which is the standard electrode potential of $Cu^{2+} + 2e^- \rightarrow Cu$, $n = 2$, and $K = [CuP_2O_7^{2-}]/([Cu^{2+}][P_2O_7^{4-}]) = 10^{6.7}$, which is the stability constant of $CuP_2O_7^{2-}$, in this equation, $E^{0'} = 0.14$ V is obtained. The value of $E^{0'}$ is shifted to the base side by about 0.2 V compared with E^0 of a simple salt bath. Moreover, because the concentration of $CuP_2O_7^{2-}$ is considerably low compared with that of the main component $Cu(P_2O_7)_2^{6-}$, the value of $E^{0'}$ is further shifted to the base side, and the Cu deposition will be inhibited. Thus, even in the system where alloy deposition is difficult due to the difference in the equilibrium potential, the alloy deposition can be accomplished by making the equilibrium potentials close as mentioned above.

The possibility of the alloy deposition can be forecast to some extent by comparing the equilibrium potential in the deposition of the individual alloy components. However, because the equilibrium potential describes the static condition where the electrical current does not appear to flow, it is insufficient to explain the dynamic reaction where the transfer of the ion, the separation of the ligand, the reductive reaction, and the electrical current participate. The current–potential curve is a valuable method for examining the deposition mechanism because we can observe the dynamic status by the method. The types of the individual concurrent electrode reactions, etc., can be estimated by measuring the relationship between the electrode potential and the electrical current density, the deposition potential when the deposition of the film starts, the film composition, and the bath composition change, etc., during the electrolysis.

Concerning the alloy plating, the deposition mechanism of the alloy plating and the formation mechanism of the alloy film can be elucidated by measuring the current–potential curves of the actual plating bath and each test bath in which the individual alloy component is separately included and then comparing the data. The current–potential curve is a curve in which the relationship between the potential and the electrical current of the sample electrode (working electrode) is plotted; it is also called the polarization curve because it expresses the deviation from the equilibrium electrode potential when the electrical current is passed, that is, the polarization. Various reactions that occur in the electrode, including the side reaction, can be examined by analyzing this curve. The current–potential curve can be measured with a potentiostat. In this apparatus, the potential of the working electrode is measured with a built-in electrometer as the potential difference between the reference electrode, which is the basis of the potential, when direct current is flowing through it. The potentiostat can supply and control the energy for the electrochemical reaction in the form of the electrical potential and can measure the reaction rate by the electrical current with high sensitivity.

Outlines of the current–potential curves of the deposition reaction of Cu and Zn from the simple salt bath and the cyanide bath, which is a complex bath, are shown in Fig. 4.2 as examples. For the simple salt bath (a), the current–potential curves of Cu and Zn are mutually greatly separated, and it is understood that there is no possibility of alloy deposition. On the other hand, in the case of the cyanide bath (b), both the current–potential curves of Cu and Zn are greatly shifted to the base side, and both curves approach. Although only Cu is deposited at the potential E_1 for (b), the deposition of Zn takes place at the potential baser than E_2, and alloy deposition is possible. Moreover, the alloy film of a composition close to brass is obtained at potential E_3.

Thus, even in the system in which alloy deposition is difficult due to the separation of the standard electrode potential, alloy deposition becomes possible by approaching the current–potential curves as mentioned above.

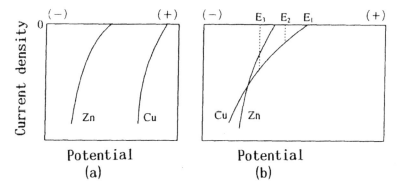

Figure 4.2 Schematic current–potential curves of deposition reaction of Cu and Zn from (a) simple salt bath and (b) cyanide bath.

Various complex baths of the cyanide bath and the pyrophosphate bath, etc., have been used for a long time for such a method in alloy plating. However, the use of the complex bath, which is composed of a strong complexing agent such as cyanide with strong toxicity, or pyrophosphoric acid with attendant difficulty in wastewater treatment, is undesirable from the viewpoint of the working environment and environmental pollution.

4.1.3 Alloy Plating Utilizing Surface Adsorption of a Surfactant

In the alloy plating of metals with greatly different equilibrium potentials, the method of adjusting the deposition potential has conventionally been adopted using a strong complexing agent as shown in Fig. 4.2(b). However, environmental problems are important issues even in the industry related to surface treatment. Currently, the creation of materials based on global environmental protection is required, and the application of the complex bath is undesirable from the viewpoint of the working environment. Moreover, there is a possibility of incurring new environmental pollution because the

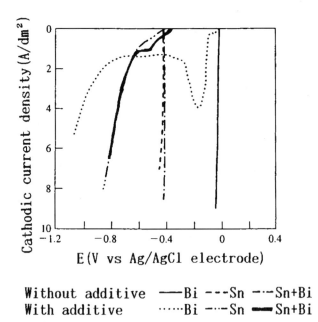

Figure 4.3 Effect of additive (surfactant) on current–potential curves in Sn–Bi alloy plating from methanesulfonate bath.

wastewater treatment is difficult. Consequently, it is necessary to construct a new lead-free solder plating process where the deposition potential is controlled without using a strong complexing agent.

Figure 4.3 shows the effect of additive (surfactant) on the current–potential curves in Sn–Bi alloy plating from the methanesulfonate bath, which is a simple salt bath. Because Bi deposits from about -0.08 V in the electrolytic bath without additive and the deposition of Sn starts from about -0.4 V, the alloy film cannot be obtained in a practicable electrical current density range. However, the deposition of Sn and Bi is inhibited along with the addition of the surfactant. In particular, inhibition of the deposition of Bi is remarkable, and the deposition of the Sn–Bi alloy becomes possible in a potential range nobler than about -0.6 V.

The phenomenon in which the surfactant is monomolecularly adsorbed on the surface of the electrode and by which the deposition reaction of the metal is inhibited is also reported in the deposition of Cu from the acidic copper sulfate bath used for manufacturing large-scale integrated (LSI) circuit interconnection and in others [6].

4.2   Sn PLATING

Although the crystal structure of Sn is tetragonal β-Sn at normal temperature, it transforms into a diamond-type, cubic α-Sn at low temperature. Electrolytically deposited Sn is deposited in the form of β-Sn even when the process is carried out below the transformation temperature, although its transformation temperature is $13\,^{\circ}\text{C}$. The density of β-Sn is 7.29 g/cm^3, the specific resistance is $11\ \mu\Omega$ cm, and the contact resistance is slightly inferior to that of gold for bright Sn plating. Moreover, because the melting point is as low as $231.8\,^{\circ}\text{C}$ and the hardness of the plating film is extremely soft as HV10 for mat plating and about HV20 for bright plating, Sn plating has been used for a long time as a surface treatment for the contact parts, the solder bonding parts, and the connectors of electronic parts. However, the needle-shaped single crystals of Sn, which are called whiskers as shown in Fig. 4.4, grow from the Sn-plated film on materials such as Cu and brass. These whiskers grow up to 1–2 μm in diameter and a length of about 1000 times the diameter, and there is a possibility of forming short circuits in an electrical circuit. This is the reason why Sn–Pb alloy plating is currently applied.

4.2.1   Plating Bath

The Sn plating bath generally being used can be roughly divided into an acidic bath, which uses a divalent tin salt as shown in Table 4.1, and an alkaline bath, which uses a four-valent stannate. Currently, the acidic bath is mainly being

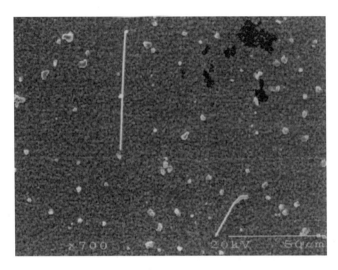

Figure 4.4 Whiskers grow from Sn plated film on Cu substrate aged for 3 weeks at
room temperature.

used because an excellent additive for the acidic bath with a fast deposition
rate was developed, although the alkaline bath was formerly the major
industrial choice.

The sulfate bath, the methanesulfonate bath, and the fluoroborate bath
are known as acidic baths. The sulfate bath and the methanesulfonate bath
are currently being used because the wastewater treatment of the fluoroborate
bath is difficult. In these acidic baths, large amounts of sulfuric acid and
methanesulfonic acid are added to prevent the hydrolysis of Sn^{2+} ion and to
secure the throwing power by increasing the electrical conductivity of the
bath. Recently, the method for depressing spongy and dendritic deposition by

Table 4.1 Typical Sn Plating Baths

Sulfate bath		Methanesulfonate bath		Stannate bath	
$SnSO_4$	40 g/L	$(CH_3SO_3)_2Sn$	60 g/L	K_2SnO_3	200 g/L
H_2SO_4	60 g/L	CH_3SO_3H	120 g/L	KOH	20 g/L
Cresol sulfonate	40 g/L	Surfactant	Proper	Temperature	70°C
Gelatin	2 g/L	Temperature	40°C		
β-Naphthol	1 g/L				
Temperature	20°C				

the adsorption of a nonionic surfactant at the Sn crystal growth point has been adopted, although gelatin and naphthol were formerly added for smoothing the plated film. Moreover, the coaddition of a reaction product of amine and aldehyde, an aromatic carbonyl compound, and a nonionic surfactant are effective for obtaining a bright plating.

4.2.2 Features of Sn Plating Film

The properties of the electrolytically deposited Sn film are as follows: The crystal structure is β-Sn, the melting point is as low as 231.8°C, and the hardness of the plating film is extremely soft. Moreover, the specific resistance of the bright plated film is 11 μΩ cm, and its contact resistance is slightly inferior to that of gold. Based on these characteristics, Sn plating has been used as a surface treatment for contact parts, solder bonding parts, and connectors for electronic parts. However, whiskers shown in Fig. 4.4 grow from the Sn-plated film on materials such as Cu and brass, and there is a possibility of creating a short circuit when applied to a minute electrical circuit. This is the reason why Sn–Pb plating has been applied.

There are many obscure points in the mechanism of the whisker growth from the electrolytically deposited Sn film for which the surface oxidation theory [7], the occluded hydrogen theory [8], the internal stress theory [9], and the intermetallic compound (which is formed with a Cu substrate) theory [10], etc., have been proposed. However, the factors in the whisker growth confirmed (proved) until now are as follows, and the application of Sn plating, which has taken measures to prevent these factors, is being examined.

1. The whiskers are not created immediately after plating but are created after an induction time of several days or several weeks. In particular, whiskers can be created and grow within a short time for the bright Sn-plated film, which has a large compressive stress.
2. The whiskers are created and grow from an Sn plated film on Cu or brass material even under a static condition in appearance during aging at room temperature. However, when an intermediate Ni or Ag plating is given to the Cu material, the creation and growth of the whiskers are remarkably inhibited.

Based on the abovementioned viewpoint, the adoption of bright Sn plating in which a big compressive stress exists should be avoided when a pure Sn plating film is applied for the surface treatment of electronic parts suitable for lead-free solder bonding. It is effective for suppressing the whiskers in plating onto the Cu base material to avoid direct Sn plating and to adopt an intermediate Ni plating [11] or Ag plating [12] process. The Sn plating process is extremely advantageous from the viewpoint of controlling the plating bath and the plating film composition although it has some disadvantages such as a somewhat higher melting point and requiring an intermediate plating process when compared with the following Sn-based alloy plating.

4.3 Sn–Ag PLATING

The eutectic composition of the Sn–Ag alloy is Sn 96.5 wt.% (96.2 mol%), and the eutectic temperature is 221 °C at an Ag content of 3.5 wt.%. Because it excels in bonding strength and heat-resistant fatigue characteristics, applications to automotive parts, portable electronic devices such as notebook-sized personal computers, and high-reliability parts have been examined, although the soldering temperature is higher compared to the Sn–Pb eutectic solder and the material cost is double.

The standard electrode potential of Ag $(Ag^+ + e^- \rightarrow Ag, E^0 = 0.763$ V) is extremely noble compared with that of Sn $(Sn^{2+} + 2e^- \rightarrow Sn, E^0 = -0.138$ V). Consequently, the plating bath containing a strong complexing agent for Ag^+ ion is indispensable in Sn–Ag alloy plating. Currently, an Sn (II) pyrophosphate bath in which Ag^+ ions are changed to complex ions with a large amount of KI is primarily being researched and developed [13–15]. However, even when this complex bath is used, the preceding deposition of Ag and the substitution deposition of Ag on the anode and plating film can take place. Moreover, when an insoluble anode is used, there are basically some problems in the temporal stability of the bath such as the formation of I_2, which is an oxidant, and in the oxidation of Sn^{2+} ion, etc. In order to industrialize this bath system, it is indispensable to use an electrolytic cell that has a diaphragm preventing the anodic oxidation of I^- or to examine a complexing agent, which takes the place of KI, for Ag^+ [16]. Moreover, it is necessary to consider the relation to the phosphorus restriction in addition to the wastewater treatment of Sn^{2+} and Ag^+ ions.

4.3.1 Sn–Ag Plating from Methanesulfonate Bath

The results of the examination concerning the Sn–Ag eutectic alloy plating from the methanesulfonate bath shown in Table 4.2 will be described in the following [17]. The effect of the additive on the Ag content in the film is shown in Fig. 4.5. The preceding deposition of Ag occurred from the electrolytic bath without an additive. Moreover, the electrical current density dependency of the Ag content was large even with the single addition of L-cysteine or polyoxyethylene-α-naphthol (POEN, mol number of ethylene oxide polymerized = 10), and the preceding deposition of Ag could not be suppressed. On the contrary, the deposition of Ag was hardly observed from the bath where L-cysteine and POEN had been added together over a wide range of

Table 4.2 Bath Composition and Plating
Condition of Methanesulfonate Bath for
Sn–Ag Eutectic Alloy Plating

$(CH_3SO_3)_2Sn$	0.192 mol/L
CH_3SO_3Ag	0.008 mol/L
CH_3SO_3H	2.0 mol/L
L-Cysteine	0.04 mol/L
2,2-Dithiodianiline	0.002 mol/L
POEN	3 g/L
Temperature	25°C

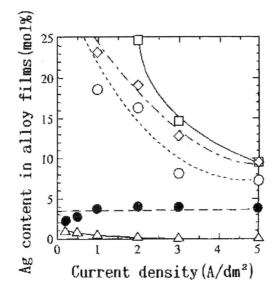

☐ Without additive
○ L-Cysteine 0.04mol/L
◇ POEN 3g/L
△ L-Cysteine 0.04mol/L+POEN 3g/L
● L-Cysteine 0.04mol/L+POEN 3g/L
  +2,2·-Dithiodianiline 0.002mol/L

Figure 4.5 Effect of additive on Ag content in Sn–Ag alloy films and its current
density dependency.

electrical current densities. From the bath where 2,2′-dithiodianiline was added, a plated film having a composition close to the eutectic composition was obtained over a wide electrical current density range of 0.2 – 5 A/dm^2 where the electrical current density dependency of the Ag content was small. Moreover, the oxidation of L-cysteine was inhibited by adding 2,2′-dithiodianiline, and the stability of the bath was improved. The addition of POEN was effective for making the film smooth and dense. Because this bath is a simple salt bath, a conventional neutralizing sedimentation treatment can be applied in the wastewater treatment.

4.3.2 Melting Characteristics and Crystal Structure of Sn–Ag Film

Figure 4.6 shows the differential scanning calorimetry (DSC) curve of the Sn–Ag alloy film deposited from the basic bath shown in Table 4.2 at an electrical current density of 2 A/dm^2 and which has a eutectic composition and a thickness of 20 μm. The isolated film (it was mechanically peeled off the stainless steel substrate) and the adhered film on the Ni intermediate layer

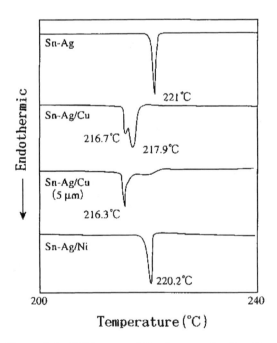

Figure 4.6 DSC curve of Sn–Ag eutectic alloy film.

melted at about 221 °C, which is the eutectic temperature of the Sn–Ag alloy. However, the film on the Cu substrate melted at 216.3°C when the coating thickness was as thin as 5 μm, and when the film thickness was thick, two endothermic peaks at 216.7° and 217.9°C were observed. It is known that the eutectic point shifts when a small amount of Cu is added to the Sn–Ag alloy; a eutectic point appears at the composition of Sn–Ag–Cu and the melting point decreases to 217°C. The melting point of the film on the above-described Cu substrate is almost equal to the eutectic temperature of the Sn–Ag–Cu alloy.

Figure 4.7 shows the results of examining the structural change according to the heat treatment (heated for 2 hr at 180°C, which is lower than the eutectic temperature, and the reflowing for 10 min at 250°C) of the film of the Sn–Ag eutectic composition on the Cu substrate by the X-ray diffraction (XRD) method. Because the diffraction peaks of the β-Sn phase and that attributed to the ε phase (Ag$_3$Sn) in the vicinity of 2θ = 40° were observed in the sample before the heat treatment, it is considered that the film of this Sn–Ag eutectic composition has a eutectic structure of the β-Sn phase and the ε phase (Ag$_3$Sn). In the sample after the heat treatment for 2 hr at 180°C, the diffraction peaks that were attributed to Cu$_6$Sn$_5$ formed by the mutual diffusion of the Cu substrate and Sn were observed in addition to the β-Sn phase and the ε phase (Ag$_3$Sn). Moreover, in the sample that is reflowed at 250°C, the diffraction peaks of the ε phase (Ag$_3$Sn) and Cu$_6$Sn$_5$ became sharper, and the growth of these crystals was suggested. The reason for the

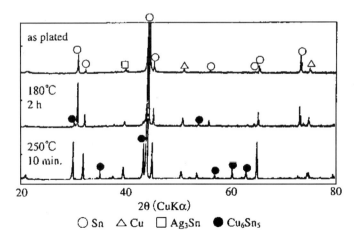

Figure 4.7 Changes in XRD pattern of Sn–Ag eutectic alloy film on Cu substrate according to heat treatment.

decrease in the melting point of the film on the Cu substrate to about 217°C is attributed to the formation of the Sn–Ag–Cu ternary alloy by mutual diffusion with the Cu substrate.

Sn–Ag eutectic alloy plating excels in bonding strength and heat-resistant fatigue characteristics, and there is no problem with the characteristics, although the material cost increases to about twice that of the Sn–Pb eutectic alloy. However, the stability of the bath and the electrical current density dependency on the film composition are particularly important in Sn–Ag alloy plating where the standard electrode potential is greatly different, and attention must be sufficiently paid to the electrochemical properties of these metals also other than the plating process.

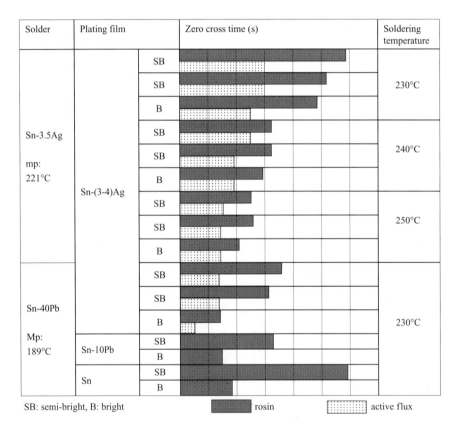

Figure 4.8 Solderability of Sn–Ag eutectic alloy film in the combination of solder material, soldering temperature, and flux.

4.3.3 Solderability and Whiskers of Sn–Ag Film

The solderabilites of the Sn–Ag eutectic alloy film were evaluated by the meniscograph method. Fig. 4.8 shows the zero crossing time in the combination of the solder material, the soldering temperature, and the flux. In the Sn–Ag eutectic alloy plated film, the zero crossing time of the bright plating film was somewhat smaller than that of the semibright plating film. Such a relation was observed in the Sn plating and the Sn–Pb alloy plating, and there is a possibility of dependency on the surface morphology. In each Sn–Ag eutectic alloy plated film, the higher the temperature of the soldering bath, the smaller the zero crossing time. In the combination of the Sn–Ag eutectic solder and the Sn–Ag eutectic alloy plated film, excellent solder wettability was observed when the soldering temperature was raised to 250°C. Naturally, the solder wettability was considerably improved by using the activated flux. From these results, we can conclude that bonding at 240°C or higher is preferable for the combination of the Sn–Ag eutectic solder and the Sn–Ag eutectic plated film.

During the aging at 50°C for 1 year, needle-shaped whiskers and nodules were not observed on the Sn–Ag eutectic alloy film on the Cu substrate. In the sample before the heat treatment, the ε phase (Ag_3Sn) was uniformly distributed in the β-Sn phase matrix, and the formation of a reaction layer was not observed at the interface of the Cu substrate and the film. However, when the sample was heat-treated for 12 hr at 180°C, which is considerably lower than the melting point, the diffusion of Ag into the Cu substrate was observed by energy-dispersive spectroscopy (EDS), and this was suggested to inhibit the formation of the Sn–Cu intermetallic compound.

4.4 Sn–Bi PLATING

The eutectic composition of the Sn–Bi alloy is tin 40 wt.% (54 mol%), and the eutectic point is as low as 140°C. The Sn–Bi eutectic alloy (tensile strength 9 kg/mm^2, elongation 25%) has a large tensile strength but a small elongation compared with the Sn–Pb eutectic alloy (tensile strength 7 kg/mm^2, elongation 55%). Accordingly, it lacks the ability to follow mechanical distortion. However, because such a disadvantage is reduced by lowering the Bi content, measures using films containing several percent Bi are being examined.

In Sn–Bi alloy plating, the preceding deposition of Bi takes place in a simple salt bath, which involves easy wastewater treatment, such as the sulfate bath, the methanesulfonate bath, etc., because the standard electrode potential of Bi ($Bi^{3+} + 3e^- \rightarrow Bi$, $E^0 = 0.317$ V) is considerably nobler compared with that of Sn, although it is not as noble as that of Sn–Ag alloy plating. Therefore, it is necessary to decrease the Sn^{2+}/Bi^{3+} ion concentration ratio in the bath below that of the intended Sn/Bi film composition, and it is difficult to strictly control the composition of the plating film. In addition, although the substitution deposition of Bi on the anode and the plating film can occur, there is a sufficient possibility of dealing with the problem technically.

4.4.1   Sn–Bi Plating from Methanesulfonate Bath

Table 4.3 shows a typical Sn–Bi alloy plating bath with the methanesulfonate bath [18]. As is clearly shown by the cathodic polarization curve in Fig. 4.3, the preceding deposition of Bi at a noble potential is inhibited by addition of a surface active agent, and the deposition of the Sn–Bi alloy becomes possible at about −0.6 V. However, the preceding deposition of Bi can occur in a low electrical current density range, and the deposition ratio of Bi decreases as the current density increases.

Sn–Bi alloy plating has been examined for about 20 years as a low-melting-point solder plating before Sn–Bi alloy plating was treated as an important subject. Because many of the lead-free solders being examined at present include Bi as a component for decreasing the melting point, Sn–Bi alloy plating can be one of the candidates of the corresponding plating film.

4.4.2   Melting Characteristics and Crystal Structure of Sn–Bi Film

Figure 4.9 shows the DSC curve of the Sn–3Bi alloy film, which was deposited from the basic bath indicated in Table 4.3 at an electrical current density of 2 A/dm^2, and has a eutectic composition and a thickness of 20 μm. The Sn–3Bi alloy film, which was mechanically peeled off the stainless steel substrate, melted at about 223°C. However, the film on the Cu substrate melted at about 221°C in the case of a thin film (5 μm). This is attributed to the formation of a Sn–Bi–Cu ternary alloy.

Figure 4.10 shows the results of XRD analysis of the structural change in the Sn–3Bi alloy film on the Cu substrate caused by the heat treatment (at 180°C, which is lower than the eutectic temperature, which was then allowed

Table 4.3   Typical Sn–Bi Alloy Plating Bath with Methanesulfonate Bath

$(CH_3SO_3)_2Sn$	0.46 mol/L
$(CH_3SO_3)_3Bi$	0.035 mol/L
CH_3SO_3H	1.0 mol/L
Nonion surfactant	10 g/L
Anion surfactant	0.7 g/L
Catechol	1 g/L
Temperature	40°C

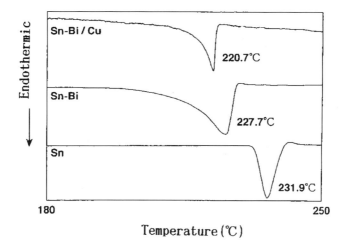

Figure 4.9 DSC curve of Sn–3Bi alloy film.

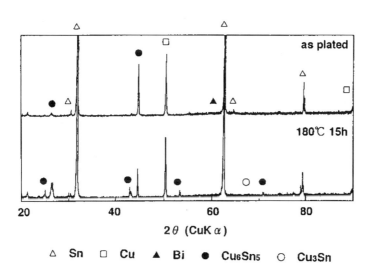

Figure 4.10 Changes in XRD pattern of Sn–3Bi alloy film on Cu substrate according to heat treatment.

to reflow for 20 min at 260°C). Because the slight diffraction peak that was attributed to the β-Sn phase and Bi was observed for the sample before the heat treatment, it is considered that this Sn–3Bi alloy film has a eutectic structure of the β-Sn phase and the Bi phase. In the sample that was heat-treated at 180°C, although the diffraction peak, which was attributed to Cu_6Sn_5 formed by the mutual diffusion of Cu substrate and Sn, was observed in addition to the β-Sn phase and the Bi phase, the formation of Cu_6Sn_5 was not observed for the sample left for 1 month at room temperature.

4.4.3 Solderability and Whiskers of Sn–Bi Film

The solderability of the Sn–3Bi alloy plated film was basically at the same level as that of the Sn–Pb alloy plated film [18,19] and exhibited excellent solderability in the combination with the Sn–Bi solder.

During the aging at 50°C for 3 months, small nodules occurred on the Sn–3Bi alloy plated film on a Cu substrate, although the generation of needle-shaped whiskers was not observed. This is because the compressive stress of the film is small and the intermetallic compound layer formed at the interface with the Cu substrate is thinner than the single Sn film. These phenomena are similar to the following phenomena: When a toothpaste tube is violently pushed, the contents having almost the same cross section as the tube outlet cross section are pushed out; on the other hand, if the tube is gradually pushed, massive contents are pushed out of the tube outlet.

4.5 Sn–Cu PLATING

The eutectic composition of the Sn–Cu alloy is Sn 98.7 mol%, which indicates a eutectic temperature of 227°C where the Cu content is only 1.3 mol%. Moreover, although the Sn–Cu eutectic alloy is inferior in oxidation resistance and requires a high bonding temperature, the bonding strength is equivalent to that of the Sn–Ag–Cu ternary alloy [20], and it is low-cost compared with the Sn–Ag alloy (equal to that of the Sn–Pb eutectic alloy). Thus it can be a promising candidate for the bonding of electronic parts used together with an oxidation inhibitor that does not negatively affect the solder joint.

In Sn–Cu alloy plating, because the standard electrode potential of Cu ($Cu^{2+} + 2e^- \rightarrow Cu$, $E^0 = 0.377$ V) is extremely noble compared with that of Sn, the cyanide bath, the pyrophosphate bath, and the borofluorate bath where the preceding deposition of Cu is inhibited have been mainly examined. However, the purpose of these baths was not to obtain the Sn–Cu eutectic alloy, but to manufacture the alloy films for speculum and bronze with much Cu content. In this paper, Sn–Cu eutectic alloy plating from the methanesulfonate bath where a thioaniline derivative and a surfactant are used as the additives will be introduced [21].

4.5.1   Sn–Cu Plating from Methanesulfonate Bath

The methanesulfonate bath shown in Table 4.4 can be applied in a conventional neutralizing sedimentation treatment in wastewater treatment. Fig. 4.11 shows the effect of the additive on the Cu content in the film and its electrical current density dependency. From the bath without the additive, the preceding deposition of Cu was observed especially in the low electrical current density range. Moreover, in the bath where either 2,2′-dithiodianiline or POEN were independently added, the electrical current density dependency of the alloy composition was extremely large, although some preceding depositions of Cu were inhibited compared with the bath without the additive. On the contrary, in the bath where 2,2′-dithiodianiline and POEN were added together, the electrical current density dependency of the alloy composition was small over a wide electrical current density range, and an alloy film, which had almost the same composition as the Sn^{2+}/Cu^{2+} ion concentration ratio in the bath, was obtained. Moreover, the current efficiency in the alloy deposition in this electrical current density range was a high value of 90–95%.

The film deposited from the bath without an additive had large particles and bare spots, and lacked smoothness. Moreover, the film that was obtained from the 2,2′-dithiodianiline single additive bath also had bare spots although the particle size became smaller. The addition of POEN was effective for smoothing and making the film dense.

4.5.2   Melting Characteristics and Crystal Structure of Sn–Cu Film

Figure 4.12 shows the DSC curves of the Sn–Cu alloy films of 20-μm thickness with various Cu contents, which were deposited at an electrical current density of 2 A/dm^2 where the molar concentration ratio of Cu^{2+} [$Cu^{2+}/(Sn^{2+} + Cu^{2+})$] in the basic bath shown in Table 4.4 was changed (however, $Sn^{2+} + Cu^{2+} = 0.2$ M). In the 0.7 mol% Cu film whose Cu content was smaller than that of the eutectic composition, endothermic peaks were observed at about 227°C, which was the eutectic temperature of the Sn–Cu alloy, and at about 230°C, which was close to the melting point of tin. Both of the films whose Cu contents were the eutectic composition (1.3 mol% Cu) and more than the eutectic composition melted at nearly the eutectic temperature.

Figure 4.13 shows the XRD patterns of the Sn–Cu alloy films with various Cu contents, which were deposited at an electrical current density of 2 A/dm^2 on the electroplated Ni films on the Cu plates. In the alloy film of the eutectic composition, because the Cu content was small, diffraction peaks of only the β-Sn phase were observed. However, in the alloy film that contained 5 mol% Cu or more, because the diffraction peaks of the η phase (Cu_6Sn_5) were observed in addition to the diffraction peaks of the β-Sn phase, it is considered that the Sn–Cu alloy film obtained from this bath has the eutectic

Table 4.4 Bath Composition and Plating
Condition of Methanesulfonate Bath for
Sn–Cu Eutectic Alloy Plating

$(CH_3SO_3)_2Sn$	0.1974 mol/L
$CuSO_4$	0.0026 mol/L
CH_3SO_3H	2.0 mol/L
2,2-Dithiodianiline	0.01 mol/L
POEN	5 g/L
Temperature	25°C

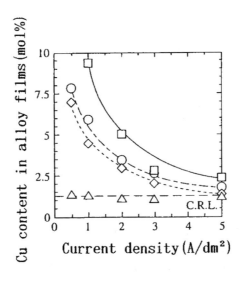

□ Without additive
◇ 2,2'-Dithiodianiline 0.01mol/L
○ POEN 3g/L
△ 2,2'-Dithiodianiline 0.01mol/L
  +POEN 5g/L

Figure 4.11 Effect of additive on Cu content in Sn–Cu alloy films and its current density dependency.

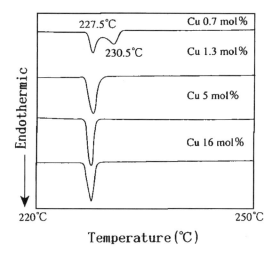

Figure 4.12 DSC curves of Sn–Cu alloy films with various Cu contents.

structure of the β-Sn phase and the η phase (Cu_6Sn_5) as expected from the phase diagram. Films having various Cu contents on a Cu substrate were allowed to reflow for 10 min at 260°C, and the changes in the crystal structure were examined. As a result, the formation of the η phase (Cu_6Sn_5) and the Cu_3Sn phase, which were considered to be formed at the interface of the Cu substrate and the alloy film due to the reflow, was confirmed regardless of the Cu content in the film. It is reported that these alloy layers have a double-layer

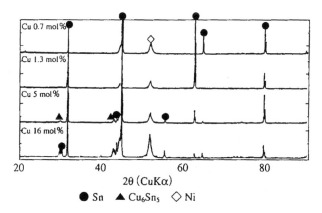

Figure 4.13 Changes in XRD patterns of Sn–Cu alloy films with various Cu contents on Cu substrate according to heat treatment.

structure of Cu_3Sn/Cu_6Sn_5, and that the Cu_3Sn layer on the Cu substrate side is comparatively thin [22].

4.5.3 Solderability and Whiskers of Sn–Cu Film

The solderabilities of the Sn–Cu alloy films with different Cu contents after the humidity test (105°C, RH 100%) were evaluated by the meniscograph method [23]. A rosin flux was used, and solders having the compositions of Sn–0.7Cu, Sn–3.5Ag–0.75Cu, and Sn–2Ag–3Bi–0.75Cu were used (temperature: 250°C). Fig. 4.14 shows the zero crossing times of the alloy films with various Cu contents for the Sn–0.7Cu solder. Almost no effect of the Cu content on the zero crossing time was observed, but some increases in the zero crossing time due to the humidity test were observed. The solderability of the Sn–Cu alloy film was basically equivalent to that of Sn single film.

Sn–Cu alloy plating films with various Cu contents were plated on the leadframes (CDA 194 copper frame and Alloy 42 Fe–Ni alloy frame) and aged for 3 months at room temperature. During the aging at room temperature, needle-shaped whiskers occurred on the Sn–Cu alloy plating film on the CDA 194 Cu frame regardless of the Cu content (Fig. 4.15). However, the formation of needle-shaped whiskers and nodules was not observed on the Alloy 42 frame. Compressive stress exists in the Sn–Cu alloy plating film. Moreover, it is hard to consider that Cu, which is the alloy component, inhibits the formation of the Sn–Cu intermetallic compound formed between the substrate. Because needle-shaped whiskers do not occur on the Alloy 42 frame, the occurrence of whiskers in this film is considered to be greatly dependent on the formation of the Sn–Cu intermetallic compound at the

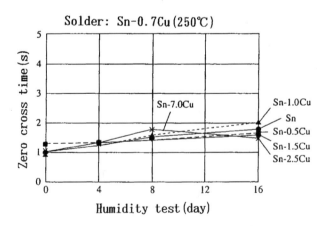

Figure 4.14 Solderability of Sn–Cu alloy films with various Cu contents after the humidity test (105°C, RH 100%).

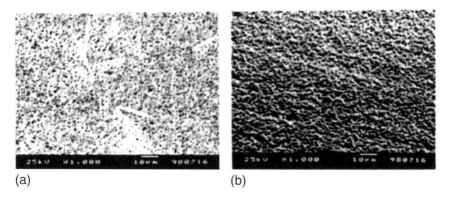

(a)   (b)

Figure 4.15 Surface morphology of Sn–Cu eutectic alloy film on leadframes aged at room temperature. Substrate: (a) CDA 194 Cu leadframe (aged 1 month) and (b) 42 alloy leadframe (aged 6 months).

interface. Thus, it is difficult to essentially prevent the occurrence of whiskers in Sn–Cu alloy plating on the Cu base material.

4.6 Sn–Zn PLATING

Concerning the characteristics demanded for the lead-free solder, it is preferable that the melting point should be close to that of the current Sn–Pb eutectic solder (183°C). The reason for this is that the working temperature of soldering, the heat-resistant temperature of the printed-wiring board, etc., are designed based on the Sn–Pb eutectic solder. The Sn–Zn eutectic alloy (Zn 9.0 wt.%), which is one of the candidates for the lead-free solder alloy, is an alloy that is expected to come into practical use because it has a lower melting point (198°C) compared with the Sn–Ag or Sn–Cu eutectic solder and is also inexpensive.

Because the standard electrode potential of Zn is extremely base in Sn–Zn alloy plating $(Zn^{2+} + 2e^- \rightarrow Zn, E^0 = -0.763$ V) compared with that of Sn, it is difficult to obtain an alloy film that has the intended composition from a simple salt bath. Therefore, the method for adjusting the deposition potential has been mainly examined using strong complexing agents such as cyanide and pyrophosphoric acid salts. However, the use of these complex baths has a problem in the working environment, and also has the possibility of creating new environmental pollution. Consequently, the development of a new plating process is required to obtain the Sn–Zn eutectic alloy without using a strong complexing agent.

Table 4.5 Bath Composition and
Plating Condition of Sulfosuccinate Bath
for Sn–Zn Eutectic Alloy Plating

$(OS_2NH_2)_2Sn$	0.193 mol/L
$ZnSO_4$	0.007 mol/L
Sulfosuccinic acid	1.0 mol/L
POEN	0.5 g/L
pH	4.5
Temperature	25°C

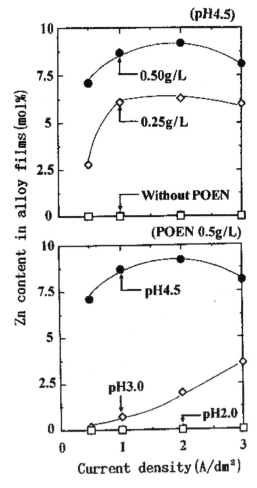

Figure 4.16 Effect of POEN addition and pH on Zn content in Sn–Zn alloy films
and its current density dependency.

In this paper, Sn–Zn eutectic alloy plating from the sulfosuccinate bath where POEN is used as an additive will be introduced along with the eutectoid mechanism of Zn [24].

4.6.1 Sn–Zn Plating from Sulfosuccinate Bath

A conventional neutralizing sedimentation treatment can be applied to the sulfosuccinate bath shown in Table 4.5 in wastewater treatment. Fig. 4.16 shows the effect of the POEN addition and the pH of the bath on the Zn content in the film and its electrical current density dependency. The factors that largely participate in the eutectoid of Zn were the POEN concentration and pH, and an alloy film, which had a small electrical current density dependency on the alloy composition and a composition almost equal to the Sn^{2+}/Zn^{2+} ion concentration ratio in the bath, was obtained under the conditions of POEN 0.5 g/L and pH 4.5. This bath was extremely stable, and even after 5 months, the sediment formation of the metastannate was not observed. The film that was deposited from the bath without an additive had large particles and bare spots, and it lacked smoothness. The addition of POEN is effective in smoothing and making the film dense.

4.6.2 Melting Characteristics and Crystal Structure of Sn–Zn Film

Figure 4.17 shows the DSC curve of the Sn–Zn eutectic alloy film, which was deposited at an electrical current density of 2 A/dm^2 from the basic bath

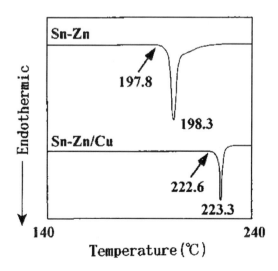

Figure 4.17 DSC curve of Sn–Zn eutectic alloy film.

shown in Table 4.5. The film that was peeled off the stainless steel substrate melted at about 198°C, which was the eutectic temperature of the Sn–Zn alloy. However, for the film on the copper substrate, an endothermic peak was observed at about 223°C, which was close to the melting point of tin. The increase in this melting point is attributed to the shift of Zn in the Sn–Zn alloy film to the Cu substrate side by diffusion due to the temperature rise during DSC. The Ni plating barrier was effective in preventing the diffusion of Zn into the Cu substrate side.

Figure 4.18 shows the changes in the crystal structure due to the heat treat ment of the Sn–Zn eutectic alloy film on the Cu substrate. In the film immediately after the electrolytic deposition, diffraction peaks of the CuZn intermetallic compound, although slight, were observed in addition to the diffraction peaks of the β-Sn phase and Zn phase. Moreover, after reflowing of the film for 20 min at 250°C, decreases in the diffraction peaks of the β-Sn and Zn phases, an increase in the diffraction peak of CuZn, and the appear-

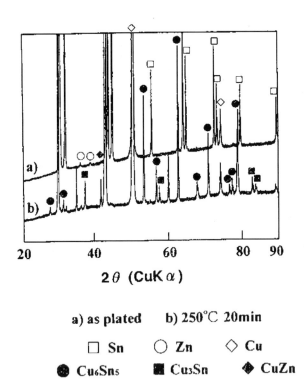

Figure 4.18 Changes in XRD patterns of Sn–Zn eutectic alloy film on Cu substrate according to heat treatment.

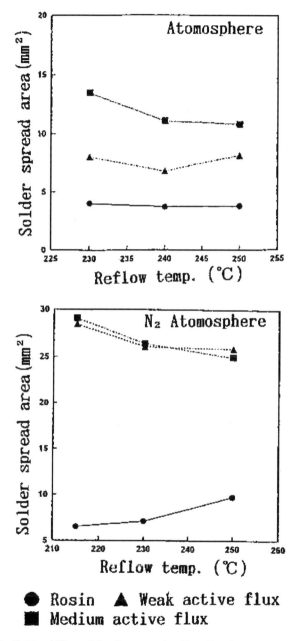

Figure 4.19 Solderability of Sn–Zn eutectic alloy film by measuring broadening area of Sn-Zn solder ball (1 mm in diameter).

ance of Cu_6Sn_5, which was the Cu–Sn intermetallic compound, were observed.

4.6.3 Solderability and Whiskers of Sn–Zn Film

Conventionally, the meniscograph method has been used as a method for evaluating the soldering capability. However, when this method is applied to the Sn–Zn eutectic solder, a large amount of dross occurs, and the evaluation is difficult. In this paper, the solderability of Sn–Zn eutectic alloy plating was evaluated by measuring the broadening area of the Sn–Zn solder ball (1 mm in diameter). Fig. 4.19 shows the relationship between the solder broadening area and the reflow temperature and flux under atmospheric conditions. The solder broadening area increased as the activity of the flux increased. It is suggested that the oxidation of the surface of Sn–Zn during the reflow occurred in this evaluation method, and the removal of the oxide layer with the nonactive flux such as rosin, etc., was insufficient. In the nitrogen atmosphere, the broadening area increased by about 1.5 times compared with atmospheric conditions, even for the rosin flux. The broadening area increased to 25 mm^2 or more by using the activated flux, and the participation of the oxide layer in the solder broadening was suggested. Currently, a flux that is effective under atmospheric conditions has been developed.

The Sn–Zn eutectic alloy film on the Cu substrate was left for 1 year at 50°C; the formation of whiskers and nodules was not observed after aging for 1 year [25].

REFERENCES

1. S Arai. Tin–silver alloy plating from non-cyanide bath. J Surf Finish Soc Jpn 49:230–234, 1998.
2. H Koyano. Development of lead-free solder plating. J Surf Finish Soc Jpn 49:235–241, 1998.
3. H Nawafune. Plating technologies adaptable to lead-free soldering. J Surf Finish Soc Jpn 49:1251–1256, 1998.
4. K Tsuji. Tin, solder and lead-free alloy plating. J Surf Finish Soc Jpn 50:155–160, 1999.
5. H Nawafune. Actualities and problems of lead-free solder plating. J Jpn Inst Electr Packag 4:276–281, 2001.
6. H Nawafune, M Awano, K Akamatsu, S Mizumoto, E Uchida, T Hosoda. Formation of ULSI copper minute wiring by copper electrodeposition process using the pre-adsorption of additive. J Surf Finish Soc Jpn 53:59–64, 2002.
7. J Eshelby. A tentative theory of metallic whisker growth. Phys Rev 91:755–757, 1953.
8. N Sabbagh. Tin whiskers—causes and remedies. Metal Finish 3:27–30, 1975.

9. S Britton. Spontaneous growth of whiskers on tin coatings—20 years of observation. Trans Inst Metal Finish 3:27–30, 1975.
10. K Tu. Interdiffusion and reaction in bimetallic Cu–Sn thin films. Acta Metall 21:347–354, 1973.
11. C Hunt. A model for solderability degradation. J Surf Mount Technol 10:32–37, 1997.
12. Y Tamura, T Hara, E Uchida, H Nawafune. Effective for suppressing whiskers by under metal coating. Proceedings of 105th Annual Conference of Surface Finishing Society of Japan, Ibaragi, 2002, pp 52.
13. S Arai, T Watanabe. Electrodeposition of Sn–Ag alloy with non-cyanide bath. Denki Kagaku 65:1097–1101, 1997.
14. S Arai, N Kaneko. Electrodeposition of Sn–Ag–Cu alloys. Denki Kagaku 65:1102–1106, 1997.
15. T Kondo, K Obata, S Masaki, K Aoki, H Nawafune. Solderability of bright Sn–Ag alloy film deposited from sulfonate bath. Proceedings of 95th Annual Conference of Surface Finishing Society of Japan, Tokyo, 1997, pp 177–178.
16. T Omi, Y Fujiwara, T Narahara, G Michinaka, H Enomoto, K Fukuda. Anode for Sn–Ag solder plating. J Electroplat Tech Assoc 10(10):9–12, 1997.
17. H Nawafune, K Shiba, S Mizumoto, T Takeuchi, K Aoki. Electrodeposition of Sn–Ag eutectic alloy from methanesulfonate bath. J Surf Finish Soc Jpn 51:1234–1238, 2000.
18. K Aoki. Sn–Bi alloy plating. Research Report of Elecroplater's Technical Association, 1998, pp 15–22.
19. I Yanada. Sn–Bi alloy plating from methanesulfonate bath. Research Report of Elecroplater's Technical Association, 1998, pp 23–32.
20. R Kawanaka, K Fujiwara, S Shigeyuki. Influence of impurities on the growth of tin whiskers. Jpn J Appl Phys 22:917, 1983.
21. H Nawafune, K Ikeda, K Siba, S Mizumoto, T Takeuchi, K Aoki. Electrodeposition of Sn–Cu alloy from methanesulfonate bath. J Surf Finish Soc Jpn 50:923–927, 1999.
22. K Suganuma, Y Nakamura. Microstructure and strength of interface between Sn–Ag eutectic solder and Cu. J Jpn Inst Metals 59:1299–1305, 1995.
23. Y Saito, K Aoki, T Takeuchi, H Nawafune. The properties of electrodeposited Sn–Cu alloy. Proceedings of 9th Microelectronics Symposium, Osaka, 1999, pp 61–64.
24. T Nakatani, K Akamatsu, H Nawafune, S Mizumoto, E Uchida, K Obata. Electrodeposition of Sn–Zn eutectic alloy from sulfosuccinate bath and properties of deposits. Proceedings of 11th Microelectronics Symposium, Osaka, 2001, pp 267-270.
25. H Nawafune, K Ikeda, S Mizumoto, T Takeuchi, K Aoki. Sn–Zn alloy electrodeposition from sulfosuccinate bath. J Surf Finish Soc Jpn 48:1007–1011, 1997.

5
Tin Whisker Discovery and Research

Yun Zhang

Cookson Electronics, Jersey City, New Jersey, U.S.A.

5.1 A HISTORICAL VIEW

According to the second edition of the *American Heritage Dictionary*, the chemistry definition of a whisker is "an extremely fine filamentary crystal that can be grown from supersaturated solutions of certain minerals and metals, and that possesses extraordinary shear strength and unusual electrical and surface properties." Although tin (Sn) whisker discussed in this chapter possesses some of the unique physical and materials characteristics, its growth phenomenon is entirely different from what was described above. In this chapter, we will specifically discuss *spontaneous* whisker growth from electroplated Sn coatings. This growth process is a solid-state phenomenon.

In 1947, Hunsicker and Kenspf [1] first discovered Sn whiskers when these two fellows at the Aluminum Research Laboratories of Aluminum Co. of America were working on improving casting properties of Al alloys for engine-bearing applications. The following is the excerpt regarding Sn whiskers from their original manuscript, Sn Trans. Q. SAE:

> An unexplained phenomenon which occurs in the solid state has been observed repeatedly in Al alloys containing tin. Microscopic examination of metallographical samples a day or more after preparation of the

polished surface has revealed thin metallic filaments extending outward from the prepared surface. These filaments appear much like minute extrusions with a thickness or diameter in the range of 1 to 4 microns and a length which varies with the time and temperature during their preparation.

The filamentary material was found to be attached in all cases to a particle of Sn–Al eutectic. However, it was observed that each small filament protrudes from only a portion of the area of an individual Sn-rich particle, that more than one filament might be associated with a single particle, and that only a relatively small percentage of the Sn-rich particles was so affected, at least in the early stages. The most remarkable example of this phenomenon is illustrated of an Al alloy-bearing insert containing 25% Sn, which had been stored at room temperature for 4 years subsequent to *machining*. A myriad of the filaments formed a deep pile projecting about 0.080 inch from the initial *machined* surfaces. A considerable quantity of this material was removed, and chemical and x-ray diffraction analyses revealed it to be principally elemental tetragonal β-Sn. The specific cause of this phenomenon remains an enigma.

This phenomenon did not receive much attention until 1952 when Herring and Galt [2] from Western Electric Bell Laboratories discovered that the mechanical strength of a Sn whisker approached the theoretical value. Further interest in whisker arose when it was discovered that the spontaneous growth of Sn whiskers in electronic equipment was causing short circuits and equipment failures [3–6]. Figure 5.1 shows the hairlike whiskers that were found on a *compressively bent* contact surface from a faulty telephone. (Photo courtesy of AT&T Bell Laboratories archives.) The risk of electrical shorts becomes a greater issue today as a result of continued miniaturization in electronic device sizes and dimensions (Figs. 5.2 and 5.3) [7]. Therefore, interest in Sn whiskers has continued as their properties have been studied [8]. Figure 5.4 clearly reflects the increased activities in this regard. During the 1960s, studies by S. M. Arnold from Bell Laboratories led to the discovery that whisker growth could be substantially repressed by small alloy additions to the finish [9]. As a result, a Sn–Pb solder finish has been adopted for use throughout the Bell System and much of the electronic industry. This was also reflected by the decline of research activities on Sn whiskers throughout the 1980s and during early 1990s (Fig. 5.4). However, since 1993, the U.S. electronics industry has recognized the inevitable move from Sn–Pb solders to lead-free solders, and had begun consortia activities to address the lead-free issues. The National Center for Manufacturing Science (NCMS), in conjunction with 11 industrial corporations, academic institutions, and national laboratories, conducted a 4-year program, known as the "Lead-Free Solder Project," to identify and to evaluate lead-free alloy

Figure 5.1 Filament tin whiskers observed on a bright tin-plated contact from a faulty telephone. (Courtesy of AT&T Bell Laboratories Archives, photo taken in 1950.)

alternatives to eutectic Sn–Pb solder. The project participants included representatives from:

- AT&T/Lucent Technologies
- Electronics Manufacturing Productivity Facility
- Ford Motor Company
- General Motors—Hughes Aircraft
- General Motors—Delco Electronics
- Hamilton Standard, Division of United Technologies Corp.
- National Institute of Standards and Technology
- Renssenlaer Polytechnic Institute
- Rockwell International Corporation

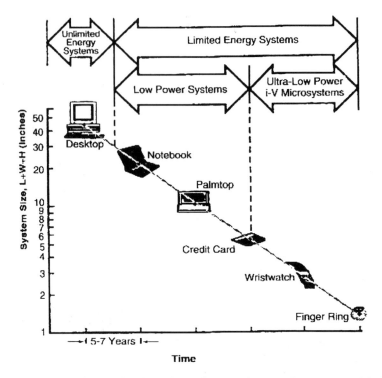

Figure 5.2 Reduction in the dimensions of personal computing systems with time. (From S Malhi, P Chatterjee. Circuits Devices Mag 10:3,13, 1994.)

- Sadia National Laboratories
- Texas Instruments Incorporated.

The project was also sponsored by the U.S. Air Force, Air Force Material Command, Wright Laboratories, and Manufacturing Technology Directorate. The Department of Defense generated US$1 million toward the US$10 million effort. The remainder of the cost came from the participating organizations.

The project goal was to determine if safe, reliable, nontoxic, and cost-effective alternatives for Pb-based solders exist for electronics manufacturing applications. The conclusions that can be made from this study are that: (a) there is no drop-in replacement for eutectic Sn–Pb solder; (b) the three alloys that showed the most promise, to the project participants, were Sn–58Bi, Sn–3.5Ag, and Sn–3.5Ag–4.8Bi; (c) each alloy evaluated has its advantages and disadvantages with respect to eutectic Sn–Pb; (d) other alloys such as Sn–2.8Ag–0.8Cu–0.5Sb show performance and marketability promise, depend-

Figure 5.3 Shrinkage in the aluminum interconnection line width in high device density IC chips. (From S Malhi, P Chatterjee. Circuits Devices Mag 10:3,13, 1994.)

ing on the usage and components being soldered; (e) although Sn–58Bi shows the most promise as replacement, known supplies of Bi will not withstand long-term use of this alloy and will cause a continuing cost increase (currently, Sn–Bi is the closest to Sn–Pb in cost); (f) alloys containing Ag will tend to increase the alloy cost significantly over Sn–Pb alloy; (g) there are no data on how lead-free solders will perform on boards that previously used a Pb-based solder product; and (h) new fluxes need to be tested with the lead-free alloys. Table 5.1 summarizes the NCMS findings.

While participating in one of the project review meetings in conjunction with the Surface Mount International (SMI) Conference in San Jose, CA in 1994, it became clear to the author that the lead-free efforts were mostly focused on soldering alloys and board finishes; not enough attention was given to component finishes. To manufacture a true lead-free product, soldering alloy, board, and component finishes would all have to be lead-free. Therefore, a concerted effort was initiated in 1994 in the author's laboratory to develop lead-free component Sn and Sn alloy finishes and to further understand the whisker growth phenomenon. In 1996, the first of a series on the subject of lead-free component finishes and whisker growth phenomenon [10] was published from the author's laboratory. In this publication, it was shown that a large, well-polygonized Sn deposit could be obtained consistently from a proprietary chemistry and was resistant to whisker formation. Detailed chemistry and materials characterizations were

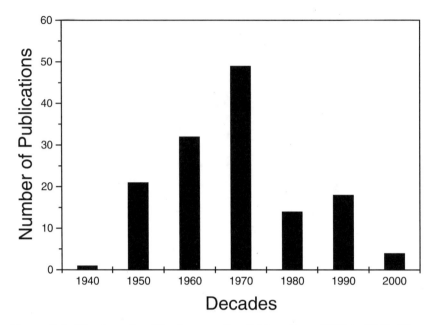

Figure 5.4 Number of publications on tin whiskers from 1947 to 2000. (From *Chemical Abstracts.*)

provided to illustrate the desirable features of the pure Sn finish. The grain size argument was further demonstrated and articulated in a subsequent patent in the context of whisker reduction [11]. Since then, there has been a growing interest on using pure Sn as a board and component finish in the electronics industry because it is the most logical and simplest replacement for Sn–Pb finishes [12,13].

At the same time, there is another school of thought. Because alloying Sn with Pb has been shown to "prevent" whisker formation [9], binary alloys of Sn–Ag, Sn–Cu, and Sn–Bi would have been a safer choice. The aerospace industry and military agencies as well as some governmental laboratories have supported this view. In recent years, some large electroplating chemistry suppliers have made statements that the binary Sn alloy deposits were whisker-free [14] whereas others have reported incidences of Sn whiskers growing from Sn alloys such as Sn–Pb and Sn–Bi [15,16]. Attempts to validate or to further understand the whisker behavior for Sn alloys were hindered by the lack of commercially viable, stable Sn alloy plating processes because of the intrinsic difficulties in codepositing these Sn alloys.

In 1999, the author received a research grant in the amount of US$1.27 million from the U.S. Department of Commence [17] to develop lead-free

Table 5.1 Summary of NCMS Findings of the Lead-Free Solder Project Ended in 1997

Alloys	Liquidus and solidus temperature	Industry	Evidence for recommendation
Sn–58Bi	139°C (eutectic)	Consumer Electronics Telecommunications	Simple two-component eutectic alloy but low eutectic temperature restricts maximum use temperature. Surface mount technology: better fatigue life than eutectic Sn–Pb for both thermal cycling ranges; less fatigue damage than eutectic Sn–Pb seen in surface mount cross sections Through-hole technology: mixed results with fatigue life for CPGA-84 better than eutectic Sn–Pb; for CDIP-20, worse than eutectic Sn–Pb
Sn–3.5Ag–4.8Bi	210–205°C	Consumer Electronics Telecommunications Aerospace Automotive	Surface mount technology: longer fatigue life than eutectic Sn–Pb at 0°C to 100°C; no failures in 1206 resistors up to 6,673 cycles; fatigue life damage than eutectic equivalent to eutectic Sn–Pb at −55°C to +125°C; less fatigue damage than eutectic Sn–Pb seen in surface mount cross sections Through-hole technology: most joints show failure by fillet lifting
Sn–3.5Ag	221°C (eutectic)	Consumer Electronics Telecommunications Aerospace Automotive	Simple two-component eutectic alloy Surface mount technology: fatigue life equivalent to eutectic Sn–Pb at 0°C to 100°C; worse than Sn–Pb at −55°C to +125°C Through-hole technology: less susceptible than other high-Sn solders to fillet lifting but results with Sn–2.6Ag–0.8Cu–0.5Sb indicate that through-hole reliability still may be compromised

solder electroplating technologies under the Advanced Technology Program, cofinanced by Lucent Technologies Electroplating Chemicals and Services (EC&S). Specifically, the objective was to develop Sn–Bi, Sn–Ag, and Sn–Cu electroplating processes by understanding the physical growth mechanisms and chemical reactions utilizing a novel technique known as spectroelectrochemistry. As a result of this funding, in-depth work became possible in the author's laboratory. Some of the results are included in this chapter.

5.2 WHISKER DEFINITION

As mentioned earlier, the dictionary definition of whisker does not apply to the phenomenon described herein. As the name implies, Sn whiskers are hairlike single crystalline structures of Sn that grow outward from electroplated films. This definition, or a slight variation of it, has been utilized to define what whisker is since its discovery. As we shall see later, this definition is too narrow.

Before we introduce the new whisker definition, it helps to rule out the cases that are not Sn whiskers. These include Sn dendrites, Sn needles, and Sn pest.

5.2.1 Sn Dendrites

Sn dendrites are Sn crystallizing in a branching or treelike manner. Fig. 5.5 illustrates how Sn dendrites look like. Unlike Sn whiskers, they are not single crystals but polycrystals. They form during, not after, electrodeposition. A few factors contribute to their formation, namely, lack of surfactant or wetting agent, improper metal concentration, and plating current density. When Sn dendrites occur, they are usually the result of one or a combination of the aforementioned factors. The most significant factor is the surfactant concentration. Sn dendrites can be eliminated readily by either adjusting surfactant and/or metal concentration, and/or by adjusting the plating current density.

5.2.2 Sn Needles

This type of plating defects can be easily confused with Sn whiskers because they look like needles (Fig. 5.6). In the literature, whiskers are very often described as needles, which is incorrect. As its name implies, the needle is bigger at one end and has a sharp point at the other end, whereas in the case of a Sn whisker, its single crystal diameter remains the same throughout its entire length. This criterion can be used as a rule of thumb to distinguish a Sn whisker from a Sn needle. Sn needles form during electrodeposition, not after.

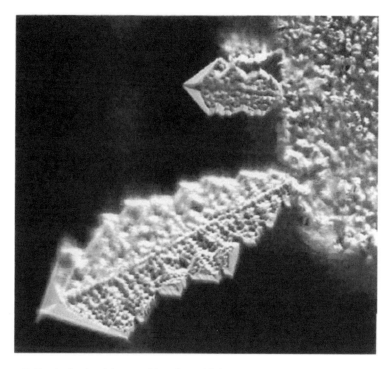

Figure 5.5 A tin dendrite resulting from high current density area (the edge of a leadframe).

A few factors contribute to their formation, namely, metal impurities, inadequate grain refiner, and/or improper metal concentrations. Sn needles can be eliminated by removing the metal impurities, and/or by adjusting the grain refiner and/or the metal concentration.

Figure 5.7 shows another example of Sn needles. This needle resulted from improper substrate pretreatment process. This phenomenon is most pronounced with Olin 151 substrate. During the plating process, Sn plated over the protruding substrate and formed a Sn needle.

5.2.3 Sn Pest

There are two allotropic forms of tin: white (β) and gray (α). The gray Sn is also known as Sn pest. Although white Sn crystallizes in the body-centered tetragonal system, gray Sn has a diamond lattice. It is considerably less dense than white tin, and is nonmetallic in appearance and properties. It is a semiconductor. The allotropic transformation occurs at $13\,^{\circ}$C and is ex-

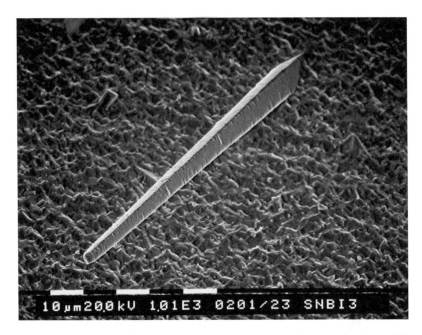

Figure 5.6 A picture of a tin needle resulting from Bi contamination. (Photo courtesy of Pascal Oberndorff from Philips Semiconductors.)

tremely slow. In fact, the existence of gray Sn was not discovered until the mid-19th century, when some Sn organ pipes in Moscow were found to have disintegrated during an exceptionally cold winter. This transformation is inhibited or prevented by incorporating a few tenths of a percent of antimony, Pb, or Bi into tin. Because Sn pest can also be suppressed by additions of small amounts of other metals, it is sometimes mistaken as Sn whisker.

5.2.4 Sn Whiskers

With the above understanding, we are now ready for a definition of Sn whiskers. Sn whiskers are Sn and Sn alloy materials that extrude from the electroplated thin Sn films after plating when the compressive stress in the Sn layer exceeds a certain threshold. They can exhibit different forms. Figure 5.8 illustrates four most commonly observed whiskers in the author's laboratory. Type 1 is called a filament. It is mostly associated with bright finishes. They can grow up to a few millimeters. Type 2 is called a nodule. It is also mostly associated with bright finishes, but normally appears as precursors to filament whiskers. The filament whiskers grow from or on top of nodular whiskers. The shape and the size of nodular whiskers vary greatly. Type 3 is called a

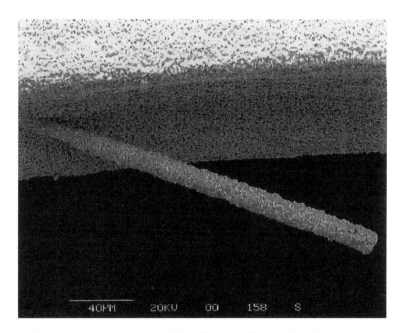

Figure 5.7 A substrate defect looking like a needle resulting from improper pre-treatment of the substrate. During the subsequent tin plating process, tin plated over the protruding needle and formed a tin needle.

column whisker. It is mostly associated with matte finishes. Its diameter is normally between 2 and 4 µm, and its length varies from a few microns up to hundreds of microns. From a practical perspective, the filament and column whiskers can be considered the same; however, they do differ in their shapes and diameters, and in what type of finishes they grow from. The diameter of a column whisker is about the average grain size of a matte finish, whereas the diameter of a filament is about the average grain size of a bright finish, which is less than 0.4 µm. In addition, the column whisker sometimes is a multi-crystal, whereas a filament whisker is a true single crystal. Type 4 is called a mound. It is mostly associated with satin bright finishes. Its diameter is slightly larger than the column whisker (ca. 3–6 µm); its length normally is limited to ca. 10 µm.

Unlike the Sn dendrites and Sn needles, Sn whiskers form after plating. The incubation time varies from a few days to a few years. Deceivingly, whisker growth does not appear to require environmental stimuli; therefore, it is often referred to as *spontaneous* whisker growth. As we shall discuss in "Key Factors Affecting Whisker Growth," environmental stimuli are always there; it is a matter of recognizing them or not.

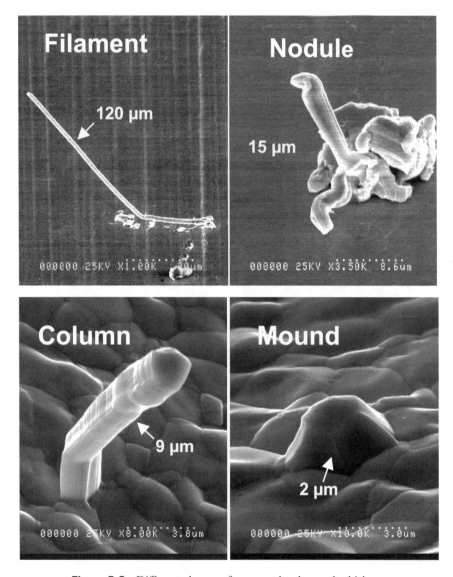

Figure 5.8 Different shapes of commonly observed whiskers.

As one can clearly see in Fig. 8, the filament whisker is only one of the four types of Sn whiskers. That is why the old definition is too narrow.

5.3 TIN WHISKER PROPERTIES

5.3.1 Chemical Composition

It is natural to assume that Sn whiskers consist of pure Sn because they originate from a pure Sn layer. Indeed, it has been the assumption since its discovery until recently. Today, we know that some whiskers consist of 100% Sn whereas others show various amounts of Cu, Zn, and oxygen, although they all grow from electroplated pure Sn coatings [23,24]. Figure 5.9 displays two cases. The filament whisker in Figure 5.9b was observed on a bright Sn coating, focused ion beam (FIB) cross section and Auger elemental mapping showed that it consisted of 100% Sn. The column whisker in Fig. 5.9a was

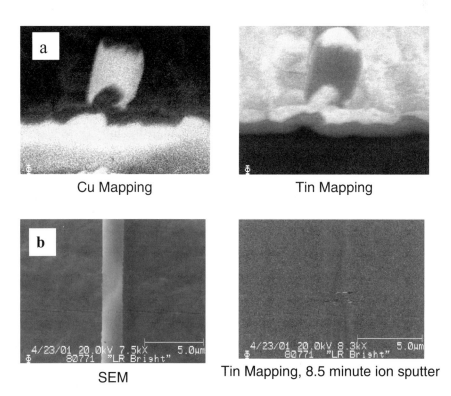

Figure 5.9 Chemical composition of whiskers: (a) a column whisker that contains a significant amount of copper, and (b) a filament whisker that contains 100% pure tin.

observed on a matte Sn finish. As shown by the Auger elemental mapping, it consisted of a Sn–Cu alloy. Given the current understanding of whisker growth mechanisms, it is not surprising that the chemical composition of a Sn whisker may not be 100% tin. Pure Sn whiskers have the crystallographical structure of white (β) Sn.

5.3.2 Physical Appearance, Shapes, and Sizes

We have shown the physical shapes of four types of whiskers in Fig. 5.8. As one can surmise, the most detrimental types are the filament and column whiskers. Even though their diameters hardly exceed 4 µm, their lengths can reach a few millimeters. The next most dangerous type is the nodule whisker especially when its base or "diameter" extends tens of microns across the surface. The mound-type whiskers do not represent any real concerns regarding bridging because they do not grow to tens of microns. Their maximum length is typically less than 10 µm. In modern microelectronics applications, the column whiskers are the most common.

Out of these four types, filaments, columns, and mounds are single crystals whereas nodules are typically polycrystals.

Close examination of filament, column, and mound whiskers reveals that they are striated (grooved) along lines parallel to their growth direction, giving the appearance of having been "extruded." Although nodule whiskers can also exhibit striations along their growth directions, their diameters are much less regular. It ranges from a few microns to 100 µm. These nodule whiskers sometimes are called "flowers" [20] and "Hershey Kiss Eruptions" by other investigators [21].

With the advent of FIB technology, it becomes possible to make cross sections of Sn whiskers. It has been confirmed by numerous researchers that whiskers are solid, not hollow [22–24]. It is important to point out here that FIB cross sections of Sn whiskers reveal that whiskers originate from inside the Sn coatings, not at the Sn surface, as suggested in Ref. 26.

5.3.3 Physical Properties

The physical properties of Sn whiskers depend on their microstructures. As mentioned earlier, Sn whiskers can be polycrystals as well as single crystals. The physical properties of polycrystalline Sn whiskers are the same as those of polycrystalline white Sn, whereas those of single crystal whiskers are rather distinct and different.

Single-crystal Sn whiskers are 1000 times stronger than ordinary Sn single crystals [2] and this strength is approaching the theoretical value. This discovery helped to resolve a problem that has confounded researchers for almost half a century: the discrepancy between the actual and the theoretical strength of solids. Most solids exhibit only a fraction of their ideal strength. This is because their single crystals are never perfect. They possess crystal defects such as dislocations, vacancies, etc. Crystals of Cu, Sn, and Al deform at stresses 100–1000 times less than those suggested by theory. The large and apparently unbridgeable gap between the real and the ideal seemed to indicate that the theoretical strengths were too high. Investigators began to ask whether the theoretical estimates might be wrong. The answer, provided by the whiskers of Sn, was definitely no. The small dimensions of these whiskers apparently allow little room for the defects that weaken larger crystals, and their strength closely approximates the prediction of theory. The growing of Sn whiskers and the study of the process by which these nearly ideal crystals grow have led to a deeper understanding of the relationship between crystal structure and many of the gross properties of solids, including their magnetic and mechanical behavior.

In addition to their strength, these tiny whiskers also exhibit exceptional electrical properties. The current-carrying capacity of a Sn whisker at a given base plate temperature approaches its maximum value as whisker approaches its melting point. Assuming cooling along the length of the whisker by conduction only, this relationship has been shown to be [26]:

$$I_{\text{melt}} = 0.076 * A/L \, [\text{mA}] \tag{5.1}$$

where A is the cross-sectional area of the whisker [in μm^2] and L is the whisker length [in cm]. For a whisker whose diameter is 2.8 μm and whose length is 0.8 mm, this value is 85 mA. The actual measured capacity of a whisker with these dimensions was 22 mA [27]. The discrepancy between the predicted and the actual value is due to idealization in the predictor equation, and the fact that area measurements of whiskers are complicated.

Based on their cross-sectional area and length, whiskers that create short circuits fall into three categories: those which sustain current flow indefinitely, those which burn out within seconds or microseconds, and those which burn out instantly. A statistically significant empirical mapping between whisker dimensions and these three categories does not exist; however, hard failures due to shorting tend to occur in high-impedance, low-current applications, whereas transient failures tend to occur in medium-power circuits drawing between 1 and 60 mA of current [28]. At ambient pressures of 0.5 Torr and below, electrical fields around the sharp tip of a whisker can accelerate electrons in the surrounding gas, resulting in a discharge capable of melting the whisker tip. When this blunt whisker makes contact with an opposing surface, a transient conducting path with a current-carrying capacity of several hundred amperes may occur. This phenomenon has been termed "vacuum metal arcing." Experiments have shown that vacuum metal arcs will not appear when there is less than 12 V of voltage across the whisker, but they may appear when the voltage exceeds 18 V. Current levels of 15 A and above have sustained vacuum metal arcs across 30 V [28]. The arcs persist until available Sn has been consumed, or until no additional current is available.

Although it has been studied extensively during the 1960s and 1970s, it is not widely known that Sn whiskers are also "one-dimensional" supercon-ductors [29–31].

5.4 WHISKER GROWTH PHENOMENON—KEY FACTORS AFFECTING WHISKER GROWTH

Whisker growth is associated with many pure metals, including Sn, Cd, Zn, Sb, and In, and less frequently with Pb, Fe, Ag, Au, Ni, and Pd [8]. It is a phenomenon most known to electroplated coatings but it also occurs with thin Sn coatings (<1 μm) resulting from electroless processes.

In this chapter, we will concentrate on whisker growth phenomenon from electrodeposited Sn coatings.

Although there is general agreement on the attributes of whiskers, especially Sn whiskers, considerable controversy exists concerning the con-ditions and reasons for growth. Indeed, much conflicting evidence has been reported in the last 30 years [22,27]. Controversy stems from the fact that the precise cause of whisker growth is not understood; consequently, potentially crucial variables have not been controlled during experimentation [32]. Aggravating the situation is the fact that spontaneous whisker growth has occurred as late as 20 years after the plating operation was performed, forcing meaningful experiments to span for decades.

Although the mechanisms by which whisker grows are not well under-stood, some key factors that influence the whisker growth have been identi-fied and have been confirmed over the last 50 years. There still exist some inconsistencies among the reported body of work. However, owing to an emerging consensus of a standard whisker test method, there is less debate on some of the key issues. In this section, a detailed account of various fac-tors that contribute to whisker growth will be presented and attempts will be made in terms of ranking their relative importance.

Before we introduce the various factors, it is important to give the readers an overview of what these factors are, their relationships in real physical space, and their interactions exemplifying as chemical and metallur-gical reactions and the consequences of these interactions.

Figure 5.10 schematically displays the relationships of these factors, which include, from top to the bottom, *surface oxide layer*, *Sn layer*, *inter-*

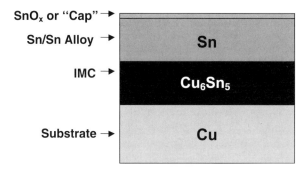

Figure 5.10 Schematic diagram of the relationships among various factors that are key to whisker formation. These factors include surface oxide layer, tin layer, intermetallic layer, and substrate layer.

metallic compound layer, and *substrate layer*. Each of these factors would influence whisker formation from the Sn layer as it will become evident later.

5.4.1 Surface Oxide Layer

Owing to the affinity of Sn toward oxygen, a thin film of surface oxide layer forms immediately after electrodeposition. Because the free energies of formation for the various Sn(II) and Sn(IV) oxide/hydroxide species are very close in value (e.g., -515.0 kJ/mol for SnO vs. -515.8 kJ/mol for SnO$_2$), the surface oxide formed is likely a mixture. The ratio of the two types is governed primarily by kinetics. Applying low-energy electron loss spectroscopy (LEELS), Bevolo et al. [33] showed that the oxide formed on both polycrystalline and single-crystal Sn from room temperature to nearly the melting temperature is continuous after one monolayer of coverage and is sharply enriched in SnO$_2$ at the outer surface. For Sn exposed to ambient environment, the same group [34] found that the outer surface was completely covered with the SnO$_2$ layer.

The amount of oxide formed on Sn has generally been found to increase with the square root of time [34,35] or logarithm of time [34,36]. The activation energy would be expected to depend on the type of Sn finish and its surface morphology, the oxidizing atmosphere, and temperature range. A systematic study of these factors has yet to be performed.

From understanding its role in whisker formation, the surface oxidation under ambient conditions is of primary concern because most of the electroplated components are stored under ambient conditions before final assembly. In their work, Britton and Bright [36] applied coulometrical electrochemical reduction to show that oxides formed on Sn and tinplate at room temperature for periods of up to 5 years are generally less than 5 nm (50 Å) thick. They observed that about 1.5 nm of oxide forms within a few minutes and grows rapidly to about 2 nm within a week, and then much more slowly to a maximum thickness of 7 nm after as long as 20 years. Britton and Bright [36] also found that the oxides on Sn grow 50% thicker when the relative humidity is very high (80%).

It has been proposed that the surface oxide layer is a necessary condition for whisker formation [37]. In addition, it is beneficial for whisker reduction because it acts as a physical barrier to prevent whiskers from protruding through. In this regard, theoretically, if a perfect surface oxide layer exists, there would be no whisker formation. In reality, the surface oxide layer is never perfect. There will always be defects such as pores and pinholes and surface corrosion products. These defect sites serve as the "weakest link" for whisker formation.

To summarize, after plating, a continuous thin film of Sn oxide on the order of a few nanometers forms. The chemical composition of the outer layer is mostly SnO_2. With increased temperature and humidity, a thicker oxide layer forms. These results support the observation that while everything else is the same, the whisker growth propensity is higher at room temperature than at room temperature but with high humidity.

5.4.2 Tin Layer

The Sn layer is the focal point of this chapter. As plated, the Sn electrodeposits can have the appearance of bright, satin bright, and matte. The atomic force microscopy (AFM) images in Fig. 5.11 illustrate how they look like three-dimensionally. As can be discerned clearly, bright deposit is much smoother compared with matte and satin bright finishes. This point is further illustrated in Fig. 5.12, a plot of gloss reflectance measured at 60° vs. surface roughness (via AFM). Of particular interest in the AFM view is the distinct presence of well-polygonized and relatively smooth surface of satin bright finish. Another way to describe the different type of finishes is to look at its grain size, shape, and codeposited carbon. Table 5.2 summarizes typical grain sizes and carbon contents of various Sn deposits.

As one would expect, the bright finish has the smallest grain size and highest carbon content, whereas it is not apparent that a stain bright finish would have the largest grain size and lowest carbon content.

We shall discuss in detail later the whisker growth behavior observed on these different finishes. Here, let us examine concisely what makes a deposit bright or matte and what chemical and process parameters determine their corresponding materials properties (e.g., grain size and carbon content) in relation to whisker growth behavior.

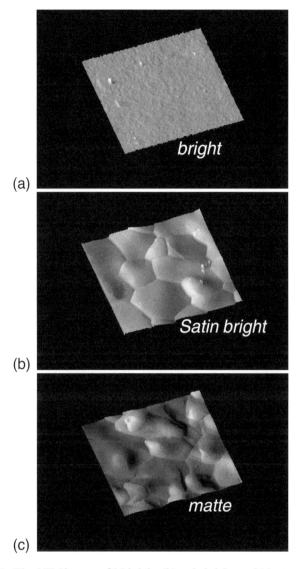

Figure 5.11   The AFM images of (a) bright, (b) satin bright, and (c) matte tin deposit.

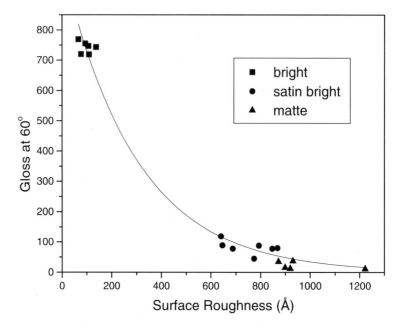

Figure 5.12 Gloss reflectance measured at 60° vs. surface roughness.

In a typical electroplating bath for Sn and Sn alloys, besides a basic electrolyte and a metal salt, there are intentionally added organic additives such as surfactants, grain refiners, and brighteners. As their names imply, surfactants are organic molecules that serve to clean and wet the substrate surface before plating. In addition, because of specific adsorption of these normally large molecules on the surface, they significantly "block" the high-energy surface sites, therefore slowing down the electrodeposition (or electro-reduction) process. Their presence in adequate amount dictates that dendritic growth does not take place. Grain refiners are classes of organic molecules that make a deposited metal surface more uniform and smooth. They are

Table 5.2 Materials Properties of Electroplated Bright, Satin Bright, and Matte Finishes

Finish	Grain size (μm)	Carbon content (%)
Bright	< 0.2	0.2
Satin Bright	~ 5	0.004
Matte	~ 1	0.01

normally smaller molecules than surfactants, and often have aromatic moieties as part of the molecule. They are also called levelers. The balance of its strength with the surfactants in a plating bath is critical. Otherwise, their effects will not be fully realized and a rough deposit or a deposit with small needles will result. Without exception, only surfactants and grain refiners are required to produce the matte or the satin bright finish. To obtain a bright finish, a brightener is employed. As its name implies, a brightener is an organic molecule that makes the deposit bright. Brighteners are small organic molecules that usually have aromatic rings in addition to other electron-rich functional groups such as a $C{=}C$ or a $C{=}O$ bond.

Nobody knows exactly how these organic molecules work on a molecular level. However, empirical experiences tell us that these organic molecules have a drastic impact on electrodeposition mechanisms. To understand this, it is useful to review the existing theories on electrodeposition mechanism. Fig. 5.13 illustrates three most popular models: 3-D island growth, 3-D island growth on top of predeposited monolayers, and 2-D layer-by-layer growth. Obviously, the 2-D layer-by-layer growth is the most desirable.

As is clearly depicted in Fig. 5.13a, when the interaction between the metal and the substrate is much weaker than the metal–metal interaction, 3-D island growth becomes the dominant mechanism.

The 3-D island growth is usually a result of not employing surfactants or employing an inadequate amount of surfactants. It is believed that nucleation initiates from the defect sites such as steps and kinks, and the nuclei grow in such a rapid rate that "islands" form, as illustrated by model 1. In practical electroplating terms, dendritic growth described previously will result from such plating bath. The deposited layer is nonadhesive and coarse due to the week metal–substrate interaction, therefore rendering it useless in practical world. Accompanying the dendrites, a portion of the surface may be exposed in severe cases, as suggested in the 3-D island growth model.

To improve the deposit quality, surfactants are added in the plating bath. It significantly slows down the electrodeposition rate as suggested by Fig. 5.13b. It is believed that surfactants preferentially adsorb on the defect sites such as atomic steps and kinks of the substrates, and they force metal ions to discharge in the atomic planes rather then steps and kinks. Because atomic planes constitute the majority of the substrate surface, there will be more nucleation sites and the initial growth will be much slower. This scenario fits model 2, which is the 3-D island growth on top of the predeposited monolayers. In this case, there is a much stronger interaction between the metal and the substrate, resulting in an adhesive layer. Also inferred in model 2 is that the first few monolayers of growth are epitaxial (i.e., the deposited metal lattice matches that of the substrate). However, because of the atomic size difference between the deposited metal and the substrate, as the layer

$\psi_{Me-S} \ll \psi_{Me-Me}$

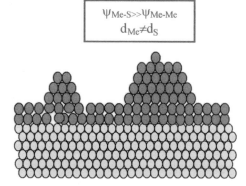

**A. "Volmer-Weber" growth model
(3-D island growth)**

$\psi_{Me-S} \gg \psi_{Me-Me}$
$d_{Me} \neq d_S$

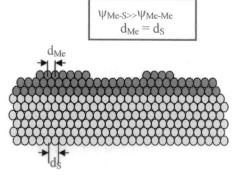

**B. "Stransky-Krastanov" growth mode
(3-D island growth on top of predeposited
monolayers)**

$\psi_{Me-S} \gg \psi_{Me-Me}$
$d_{Me} = d_S$

d_{Me}

d_S

**C. "Frank-van der Merwe" growth mode
(Layer-by-layer growth)**

Figure 5.13 Popular models of electrodeposition.

builds up, the grain size of the deposited metal increases. As a result, the surface of the finish is still relatively rough.

In a practical world, the deposits resulting from model 2 are not suitable for microelectronics applications. To further improve the smoothness of the surface, grain refiners are added in the plating bath. When properly instituted, the grain refiners can impart a much smoother surface due to overall grain size reduction. However, this level of smoothness may still not be satisfactory for some customers who prefer a mirror-bright surface because it is cosmetically more pleasing. To obtain such optical property, it is necessary to have grain sizes that are significantly smaller than the wavelengths of the visible light (0.4–0.7 μm), so that light is not scattered or absorbed but reflected from the plated surface.

As we have shown previously, the grain size can be further reduced by introducing a brightener. The ideal case when carrying out this action is to obtain a completely smooth surface, illustrated by the 2-D layer-by-layer growth model (Fig. 5.13c). In this case, in addition to a strong interaction between the metal and the substrate, the grain size of the deposited metal is similar to that of the substrate. Furthermore, the kinetics is so slow due to the strong inhibition of the brightener; the nuclei grow preferentially laminarly rather than columnarly.

Although these models are invaluable in helping formulators and electrochemists to understand electrodeposition at an empirical level, there is a need to obtain a molecular level understanding on how these additives work. This is especially true when we attempt to successfully electroplate Sn–Cu, Sn–Bi, and Sn–Ag alloys. While at Lucent Technologies EC&S, the author and her team made the first attempt toward the goal of molecular level understanding of how these organic molecules work using a commercial pure Sn electroplating chemistry [38]. We studied the chemical nature of adsorbed species by in situ surface-enhanced Raman spectroscopy (in situ SERS). In SERS experiments, only the Raman bands associated with functional groups that adsorb on the surface are enhanced. We attempted to answer the following questions: How does an organic additive adsorb on the substrate surface? What is its adsorption strength and orientation? What part of the molecule (i.e., the functional group) is in contact with the substrate surface? Is there a difference in adsorption before and during electrodeposition? How does this impact: (a) the initial stage of the grain growth; (b) subsequent layer buildup; and (c) materials properties of the finished layer? Is there a quantitative difference in adsorption strength among surfactants, grain refiners, and brighteners? The growth mechanism was studied by in situ scanning tunneling microscopy (STM). We attempted to find answers for the following questions: Where does the nucleation originate: steps and kinks, or atomic planes? How do the various organic molecules (surfactants, grain refiners, and brighteners) affect the initial nucleation and subsequent growth? How do the grain refiner and brightener work at a molecular level?

A description of these experiments and results were given elsewhere [39]. A concise summary will be provided here to demonstrate the importance of why plating bath maintenance and proper plating parameters are important in whisker reduction and prevention and, most importantly, why matte or satin bright Sn is preferred over bright tin.

In Sn and Sn/Pb plating, the most commonly used surfactants are polyethylene glycols with different numbers of ethylene oxide (EO) groups. Sometimes other surfactants are also used. In this work, a surfactant called Triton X-100, shown in Fig. 5.14, was used. It not only has ethylene oxides, which is hydrophilic, but also contains a substituted benzene ring, which is hydrophobic. Using SERS, we found that at open circuit potential (OCP) and potential equivalent to low current densities (0.2 V vs. Ag/AgCl), no Raman shift of the EO groups could be detected, suggesting that the EO groups are tilted away from the substrate surface. From the Raman spectra, it was deduced that the benzene ring and the alkyl group are in direct contact with the substrate surface. At −0.5 V, equivalent to >10 ASD in current density, the Raman bands of both the substituted benzene ring and the EO groups

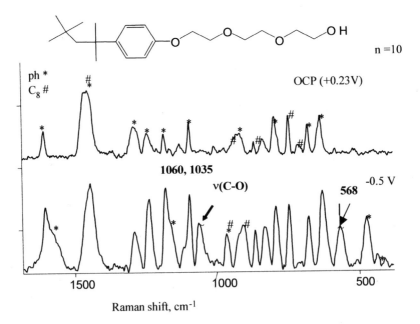

Figure 5.14 Raman spectra of Triton X-100 in a tin plating bath obtained at various potentials. The molecular structure of Triton X-100 is shown on top.

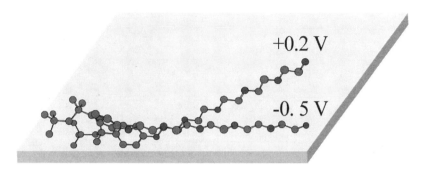

Figure 5.15 Model describing the adsorption behavior of Triton X-100 at various potentials. Light gray dots are carbon atoms and dark gray dots are oxygen atoms.

could be discerned. Based on these observations, a model shown in Fig. 5.15 was proposed. The direct consequence of specific adsorption of the organic molecules on the electrode is increase of the surface resistance, which can be measured by the double-layer capacitance. Surface resistance increase results in reduced current density at the same applied potential. This is further illustrated by Fig. 5.16. As can be surmised, the surfactant adsorption is beneficial in reducing the electrodeposition rate, resulting in a smooth surface 5 s into the deposition [Fig. 5.17b (with surfactant) compared with Fig. 5.17a

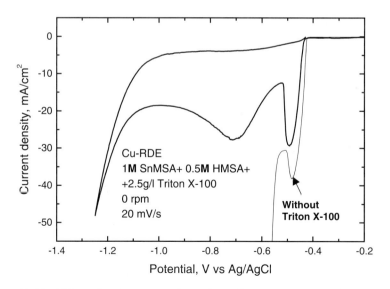

Figure 5.16 Effect of surfactant on electrodeposition rate. As seen, without Triton X-100, the deposition rate at −0.5 V is out of control.

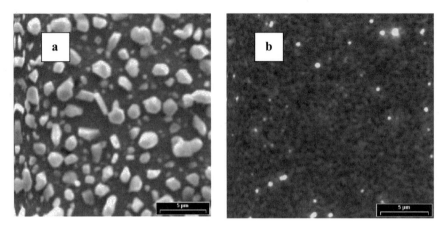

Figure 5.17 Surface morphology of the electrodeposited thin films of tin obtained 5 s after the onset of electrodeposition: (a) without surfactant Triton X-100, and (b) with surfactant Triton X-100.

(without surfactant)]. The latter develops into dendrites if longer plating time is permitted.

However, the surface obtained from this system, when it is sufficiently thick (>3 μm), is not smooth. This is because as the layer grows, the grain size of the Sn deposit grows as well. This is illustrated in Table 5.3. Results obtained from two surfactant systems are tabulated. As one can see, the deposit grain size grows into micron range during the first 20 s. To further refine the surface, we introduced a grain refiner, a compound called phenolphthalein, into the system. In the surface-enhanced Raman spectroscopic experiments, we found that most of the bands associated with Triton X-100 decreased in intensity; bands assignable to phenolphthalein became relatively much stronger, indicating that there is a competitive adsorption between Triton X-100 and phenolphthalein. The latter adsorbs stronger than the

Table 5.3 Growth of Grain Size and Roughness of the Sn Electrodeposits as a Function of Time (the First 40 s)

Time (s)	PEG		Triton X-100	
	Grain size (μm)	Roughness (μm)	Grain size (μm)	Roughness (μm)
5	0.5–1.0	0.1–0.15	0.1–0.2	0.01–0.02
20	0.75–1.5	0.15–0.4	1.0–1.5	0.1–0.2
40	1.2–2.0	0.3–0.6	2.0–7.5	0.3–0.6

former. Of most importance are the bands associated with the two phenol groups from phenolphthalein. We observed them at OCP and −0.5 V. The suggested adsorption mode of phenolphthalein shown in Fig. 5.18 agrees with the Raman data.

It is important to point out that phenolphthalein is not soluble in water, and it does not adsorb on the surface by itself. Most likely, it is "wrapped" in micelles of Triton X-100, which serves as a "carrier." The surface obtained in the presence of both Triton X-100 and phenolphthalein is much smoother than with Triton X-100 alone, and it has a satin bright appearance. Why? Exactly how does the grain refiner work? And how it is different from the surfactant in situ STM studies illustrated, that the difference lies in the first few seconds during the deposition? As shown in Fig. 5.19, instead of forming small (ca. 0.15 µm) irregular grains as in the case of surfactant alone, in the presence of the grain refiner, the Sn grains are regular (disclike) and relatively larger (ca. 0.5 µm) at the very beginning. Most important of all, these grains appear to be very flat, and disc size remains unchanged during the first 10 s. This behavior is very different from that of the surfactant. As pointed out previously, with Triton X-100 alone, grains grow significantly larger in the first 5–10 s.

The significance of this finding is the following: The grain refiners regulate the size of the grains from the beginning to the end of the deposition.

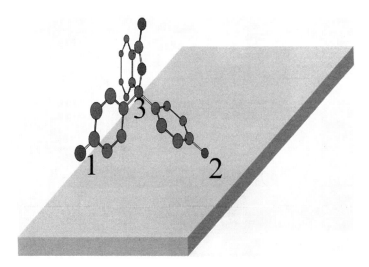

Figure 5.18 Model simulating the adsorption behavior of the grain refiner, phenolphthalein, on the substrate surface as different potentials. Light gray dots are carbon atoms and dark gray dots are oxygen atoms. This model agrees well with the in situ SERS data.

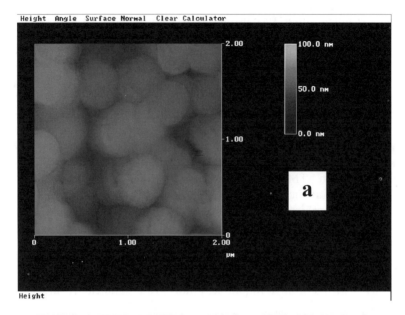

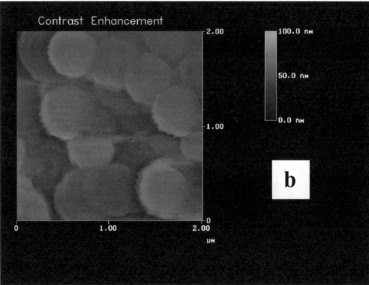

Figure 5.19 Scanning tunneling microscopy images of the grain size of tin in the presence of both Triton X-100 and phenolphthalein: (a) after 5 s, and (b) after 20 s.

Because the grains are flat discs, when they grow layer by layer, it results in a smooth surface. The 0.5-μm grain size dictates that it is not bright. It is reasonable to speculate that the brightener behaves the same way except that the grain size is smaller (≤ 0.4 μm).

We introduced a known commercial brightener, cinimonaldehyde, in the Sn bath. Preliminary results shown in Fig. 5.20 suggest that the aforementioned hypothesis holds truth. However, because a different surfactant was utilized in this experiment, we cannot compare directly the results between the grain refiner, phenolphthalein, and the brightener, cinimonaldehyde. It is important to point out that the Raman intensity of cinimonaldehyde is about a magnitude higher than those of phenolphthalein, confirming the common belief that brighteners adsorb on the substrate strongly.

Another aspect of brightener also warrants brief discussion. The brightener molecule is much smaller than both the surfactant and the grain refiner. Combined with much stronger adsorption, it can be trapped in the "gaps" among large molecules. As a result, it is codeposited with tin, resulting in high organic content in the Sn coating. Without exception, bright finishes contain much higher included carbon (ca. 0.05–0.4 wt.%) when compared with matte and satin bright finishes (ca. 0.001–0.05 wt.%).

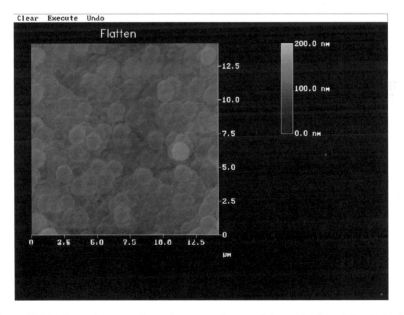

Figure 5.20 Scanning tunneling microscopy image of the grain size of the tin in the presence of polyethylene glycol and cinimonaldehyde.

5.4.3 Grain Size, Shape, and Carbon Content

Based on the discussions above, plating chemistries utilizing different organic molecules result in deposits with different appearances and properties. They can be bright, matte, or satin bright. Microscopically, there is a difference in included carbon, grain size, and grain shapes. Bright finishes have grain sizes ranging from 0.05 to 0.25 μm. The satin bright and the matter finishes, on the other hand, have grain sizes ranging from submicrons to several microns. Conventional wisdom predicts that a matte finish would have a larger grain size than a satin bright finish. We found that it is not the case. As shown in Fig. 5.21, it is evident that the matte finish has smaller grain sizes than those of the satin bright finish.

Nonetheless, how do these finishes differ with regard to whisker formation? In a systematic study utilizing these three finishes, we found a direct correlation between whisker growth propensity and the deposit grain size, shape, and carbon content. Specifically, the bright finish showed the worst growth behavior as expected; the satin bright finish performed the best. The matte finish fell close behind the satin bright finish [40].

It is important to point out that in this study, an oxygen-free annealed Cu substrate was utilized to minimize the substrate effect so the observations made with regard to whisker growth pertain mainly to the differences in the deposit properties and subsequent interfacial reactions.

In addition to the grain size, shape, and codeposited carbon, the crystallographical orientation of the Sn deposits has been observed as an important factor.

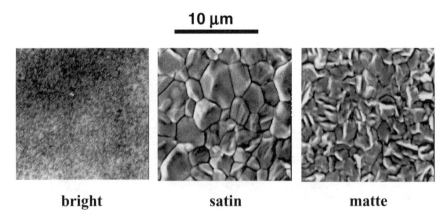

Figure 5.21 The SEM photos of three different finishes: bright, satin bright, and matte. The grain size of satin bright is larger than matte.

5.4.4 Crystallographical Orientations

Up to this day, the most studied whiskers by x-ray diffraction (XRD) have been the filaments. Crystallographical investigations have indicated that the growth directions of the filament whiskers are of the low-index faces such as $\langle 111 \rangle$, $\langle 101 \rangle$, and $\langle 100 \rangle$ [22,41,42–44], and the growth direction of the whiskers is different from the preferred orientation of the Sn deposit. Although much work needs to be carried out to systematically investigate the relationship between preferred orientations of Sn deposit and the whisker growth direction, it has been generally observed that Sn coatings with a strong preferred orientation of $\langle 220 \rangle$ have a reduced whisker propensity [45,46].

Fundamentally, it is of great interest to understand what affects the preferred orientations of the Sn deposits. Is it current density, the nature of the organic additives, the substrate, or the combination of all of the above?

It suffices to say that substrate material and its stress and orientation all influence the stress and orientation of the Sn layer. In addition, pretreatment process also plays a significant role in determining the stress and orientation of the Sn deposits. Current density has been shown to influence not only the preferred orientation but also the stress of the Sn layer. We can speculate that the nature of the organic additives also plays a role.

5.4.5 Plating Bath Conditions and Range

Plating chemistry plays a very important role in whisker growth propensity. In general, bright chemistries are prohibited in electronics applications because bright finishes have higher whisker growth propensity compared with matte finishes. Many large semiconductor manufacturers have chosen matte Sn in their microelectronics application. Most notable among them is the E3 group, a partnership formed among Infineon, ST Microelectronics, and Philips.

In a matte Sn plating bath, besides Sn salts, there are surfactants and grain refiners. As was described earlier, it is important that these components are within working range, often referred by chemical suppliers as operating windows. For instance, if surfactant concentration is too low in the bath, one may experience dendrite growth. If grain refiner concentration is too low, the plated surface may be too rough. This may lead to more than normal surface oxide formation, which may be beneficial in reducing the whisker propensity but may be less desirable for long-term solderability. On the other hand, if the grain refiner concentration becomes too high, one may experience the formation of Sn needles, which can be easily confused with Sn whiskers.

In addition, when impurities such as Cu and Fe are introduced in the plating bath, it has been reported to significantly increase whisker growth propensity [21,47]. Therefore, care needs to be taken to not introduce impurities such as Cu, Ni, Fe, and Pb in the plating bath. If these elements indeed are present in the pure Sn bath and are in the amount in excess of 50 ppm, corrective actions such as dummy plating should be taken to remove them before plating.

5.4.6 Plating Parameters

Plating parameters refer to all the factors that complete the task of plating other than bath components. They include parameters such as current density, solution agitation, parts movements, bath temperature, and cell voltage. It has been shown that when Sn deposition was carried out outside of the limiting current density where hydrogen evolution occurs, whisker growth propensity was greatly increased. It has been speculated that the included hydrogen caused high internal stress in the deposit, and was responsible for the increased whisker propensity. In addition, varying current density could also result in deposit microstructure (i.e., grain size, shape, and texture or preferred orientation) changes, which, as we have shown earlier, have direct implications with regard to whisker formation.

It is important to note that plating chemistries supplied by various chemical suppliers may differ significantly in terms of the window of operation. Sometimes the term *robustness* is often used to refer to a chemistry that is not sensitive to typical plating parameter fluctuations.

5.4.7 Interface

The word interface means a surface forming a common boundary between adjacent regions. In this case, we are specifically referring to the interface between the Sn layer and the substrate. This interface forms as a result of interfacial diffusion due to the difference in chemical energy between the two metals [48]. There are two main pathways for interfacial diffusion: diffusion through bulk and diffusion through grain boundaries. The diffusion process takes place in both directions. It has been established previously that predominantly Cu diffused into Sn for this diffusion couple. Sn diffusion into Cu is minimal.

The interfacial diffusion between Sn and Cu happens readily at normal soldering temperature, resulting in a duplex Cu–Sn intermetallic layer, comprised of Cu_6Sn_5 (η-phase) in contact with Sn and Cu_3Sn (ε-phase) in contact with Cu substrate. After a rapid initial reaction, the thickness increases as the square root of time at a given temperature [49–51]. Cu_6Sn_5 forms first [51,52] and the Cu-rich Cu_3Sn phase results from Sn depletion at the reaction site. As Sn is depleted, the Cu-rich phase, Cu_3Sn, increases in thickness at the expense of Cu_6Sn_5 phase. This Cu_3Sn phase grows at a temperature above

80°C [53]. It has been reported that both Sn and Cu are mobile in the inter-metallic phase.

Intermetallic growth rate increases exponentially with temperature [49,51,54]. The rate also depends on the microstructure of the substrate metal and whether the Sn electrodeposit is bright, matte, or reflowed [49]. At temperatures below 170°C, the growth rate is faster on hard Cu than on soft Cu substrate, and it is faster with a bright finish than a matte finish. As might be expected, the presence of steam has no effect on intermetallic growth rate [53]. This point is important when we consider an accelerated whisker test method.

Although Sn–Cu interface has been studied extensively, there is a lack of detailed understanding with regard to the following questions: (a) Does the interfacial diffusion take place faster at the grain boundaries or through the bulk? How is this phenomenon dependent on the temperature? (b) What is (are) the morphology of the intermetallic phases? Does the morphology change with temperature? (c) Is there a grain growth accompanying the in-termetallic layer thickness growth at normal annealing and soldering temperatures?

Recently, Lee and Lee [55] and Tu [56] reported their investigation of the diffusion process between a Cu substrate and an electroplated matte Sn finish. At room temperature, Tu reported an irregular intermetallic layer of Cu_6Sn_5 between Cu and Sn. Similar observations were made in the author's laboratory [18] (Fig. 5.22). In addition, we found that at room temperature and 50°C, the intermetallic compound growth takes place predominantly at Sn grain boundaries (Fig. 5.23). Direct stress measurements in the Sn layer by x-ray diffraction immediately after plating and 3 months later showed zero stress and a compressive stress of ca. 10 MPa, respectively. Apparently, this stress was built up during storage at these temperatures, and is responsible for whisker growth a few months after plating.

At elevated temperatures, the growth behavior of the intermetallic compound is different. Growth mechanism appears to be taken over by bulk diffusion. Consequently, a relatively dense and regular intermetallic layer is formed (Fig. 5.24). Accompanying the intermetallic compound thickness increase, there is significant grain growth at these temperatures. Fig. 5.25 shows the grain growth of Cu_6Sn_5 at 175°C at various time intervals. As one can see, the average grain size of Cu_6Sn_5 has grown to 1 and 4 μm after 5 and 60 min, respectively. Simultaneously, the Sn grains have grown to an average of 10 and 30 μm, judging by the "residue" Sn grain boundaries.

In addition, direct stress measurements on reflowed samples showed that a compressive stress of ca. 7 MPa in the Sn deposit measured before reflow was completely removed after reflow owing to grain growths (both Cu_6Sn_5 and Sn). Furthermore, these samples have been stress-free 18 months

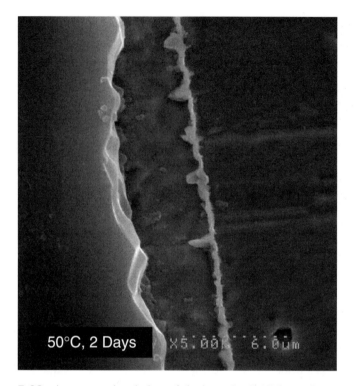

50°C, 2 Days   X5.00K   6.0μm

Figure 5.22 A cross-sectional view of the irregular IMC formation at 50°C.

after reflow, suggesting that the dense and regular intermetallic layer formed during reflow process serves as a barrier to effectively stop the additional interfacial diffusion between Sn and Cu. The stress generation source has been essentially removed.

In summary, irregular intermetallic compound formation results in compressive stress in Sn. The irregular growth behavior is associated with diffusion through Sn grain boundaries, which is dominant at ambient temperatures (room temperature and 50°C). At elevated temperatures such as 150°C and 175°C, the intermetallic growth behavior is governed by bulk diffusion, which results in a dense, regular intermetallic compound layer. At these temperatures, the stress relaxation due to grain growth largely removes any residue stress in the Sn layer, therefore effectively eliminating the danger of whiskering. No or little compressive stress will build up again because the dense and regular intermetallic layer acts effectively as a barrier layer.

To elaborate further, the key to achieving whisker freedom is to remove any compressive stress source. One effective way to do this is to introduce a

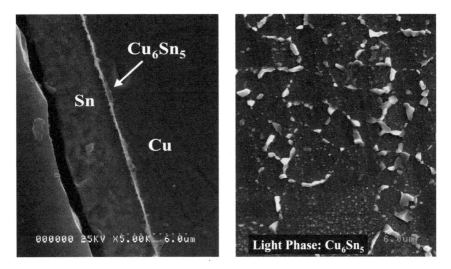

Figure 5.23 The SEM photo of the Cu_6Sn_5 phase. Left, cross-sectional view; right, top view.

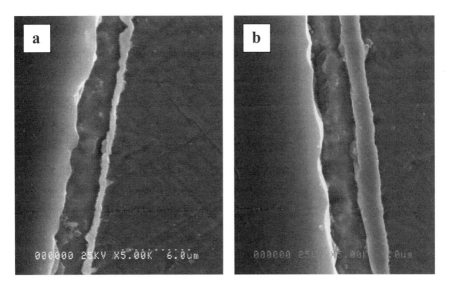

Figure 5.24 Cross-sectional views of the regular and dense IMC formation at (a) 120°C, 60 min; and (b) 175°C, 60 min.

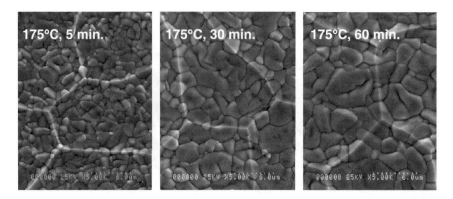

Figure 5.25 The IMC and tin grain growth at 175°C.

barrier layer such as a Ni layer in between Cu and tin. Studies have shown that a Ni layer from 0.2 to 1.5 μm is effective in eliminating whisker formation, although there are contradictory reports.

5.4.8 Substrate and Pretreatment

The most common substrates for Sn and Sn alloy plating are Cu and Cu alloys. Examples include Olin 194, 151, 7025, phosphor bronze, brass, and pure Cu. Non-Cu substrates such as alloy 42 (a Ni–Fe alloy) and steel are also utilized.

The nature of the substrate has long been recognized as having a profound effect on whiskering. It is generally observed and agreed on that among the most common substrate materials, brass has the worst effect, followed by phosphor bronze. Cu and Cu alloy, Ni, and Fe are comparatively much better. In particular, it has been reported that of the common leadframe substrates, Sn plated on alloy 42 is more whisker-resistant than on Olin 194, 151, and 7025. It has also been observed that there is a significant difference in whisker growth behavior when the leadframe material is stamped vs. etched.

Recently, it has been shown that the stress and crystal orientation of the substrate materials have significant influence on Sn electrodeposited stress and orientation [57]. In addition, certain orientations, such as $\langle 220 \rangle$ in the Sn layer, appear to be more whisker-resistant than others. At the same time, certain orientations such as $\langle 321 \rangle$ appear to be associated with the visual defects of the deposits right after plating and increased whisker propensity [41,58] during storage.

These recent observations and studies raise the following questions: (a) Is the stress in the Sn layer influenced by the substrate residue stress?

How? (b) Which has a bigger impact on the Sn deposit stress: organic additives used in the plating bath, or substrate residue stress? (c) How does the pretreatment of the leadframe materials influence the substrate stress and orientation? (d) Are orientation and stress related?

Zhang [57] and Zhang and Schetty [59] carried out a study to systematically investigate the stress and preferred orientation of the substrate material before and after pretreatment, a step before Sn plating operation. They found that the common leadframe and connector substrates all possess compressive stress before the plating step. Table 5.4 summarizes the residue stress values measured by XRD. As one can see, the substrate follows the order below in terms of decreased compressive stress: brass > Olin 7025 > Phos Bronze > Olin 194 > Olin 151.

However, when these substrate materials went through a pretreatment procedure such as descale (for leadframe plating) or cathodic clean (for connector plating), their stress states were altered. It is very important to recognize that the pretreatment process is able to introduce stress so significantly that it can become one of the most important sources for *initial* deposit stress. In addition, the same substrate possesses a different stress level when a different pretreatment process is employed. Zhang [57] found that when Olin 151 was treated with a more aggressive descale process, its stress changed from -8 ± 2 to -42 ± 2 MPa, whereas with a mild pretreatment process, its stress changed only to -16 ± 2 MPa, as shown in Table 5.4.

It is interesting to note that the preferred orientation of these Cu alloy materials does not change regardless of the pretreatment process. The preferred orientation remained as $\langle 220 \rangle$. However, different pretreatment methods do affect the preferred orientation as well as stress in the Sn layer, as suggested in Table 5.5. This finding is significant because it means that even though a chemistry can characteristically produce tensile-stressed Sn coatings on a substrate material with proper pretreatment method, if attention is not paid to the pretreatment process, under certain conditions, the same chemistry could yield a compressively stressed Sn layer.

Table 5.4 Stress Measurements of Substrates by XRD Before and After Pretreatments

Substrate	Before pretreatment (MPa)	After pretreatment (MPa)
Olin 151	-8 ± 2	-16 ± 2
Olin 194	-10 ± 2	10 ± 2
Phos bronze	-15 ± 2	-22 ± 2
Olin 7025	-56 ± 2	-32 ± 2
Brass	-73 ± 2	-62 ± 2

Table 5.5 Preferred Orientations and Stress of a Matte Sn Finish Plated on Olin 151 Substrate with Two Different Pretreatment Methods

Pretreatment method	Preferred orientation(s)	Stress (MPa)
Mild	⟨220⟩	5
Aggressive	⟨220⟩, ⟨321⟩, ⟨211⟩	−4

Nonetheless, it should be noted that very limited work has been done on this subject; more systematic investigations are needed and strongly recommended by the author.

5.4.9 External Mechanical Stress

During electronic assembly processes, there are many sources of mechanical stress exerted on the electroplated deposits. Some are intentional, and they are integral parts of the manufacturing processes. For example, after Sn plating, the leadframes go through "trim and form." This step entails that the leads are trimmed from the carrier strips, and then are formed with either a J or a gullwing bend. This bending angle is approximately 90° (Fig. 5.26). As a result of the "trim and form," part of the leadframe will be tensile-stressed, and part of the same leadframe will be compressive-stressed. On the other hand, there are unintentionally introduced mechanical stresses. For instance, improper sample handling could result in physical damages such as scratches and indentations on the plated parts.

Experimentally, mechanically applied stresses are frequently observed to result in whisker growth [18,23,27,60]. Common examples include growths

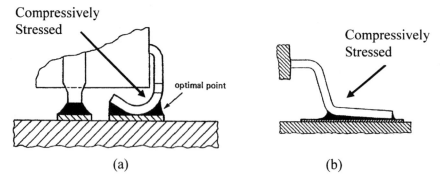

(a)   (b)

Figure 5.26 Schematic drawing illustrates J and gullwing bend.

adjacent to load-bearing connections (e.g., bolts), or at the cut-edge of stampings. In the case of Hunsicker and Kenspf [1], they polished the Sn–Al surface, which resulted in localized whisker growth days after the sample preparation. Physical damage to coatings, such as scratches or indentations, can also result in localized whisker growth. It must be said that not all investigators reported the same observations when similar experiments were performed. For instance, Dunn did not report any increase in whisker formation after application of compressive stresses to Sn-coated brass C-rings. However, it is important to point out that in Dunn's experiment, a bright Sn finish was utilized. The plated samples showed gross whisker formation days after plating even without any mechanical compressive stress, indicating that the residue stress in the Sn layer alone was high enough to induce whisker growth readily.

In summary, in addition to surface oxide layer, Sn layer, the intermetallic compound layer, and the substrate, the external mechanical stress is also one of the key factors that contribute to Sn whisker formation.

5.4.10 Thermal Stress

Thermal stress originates from the thermal expansion or contraction of materials when they are mechanically constrained. Integrated circuit (IC) fabrication generally involves bilayers or multilayers of different materials. The simplest structure is a bilayer structure, which consists of a surface film and a substrate. Internal stress will develop at the film–substrate interface because of thermal mismatch between the film and substrate at the film formation and device service temperatures. During the lifetime of an electronic device, it experiences temperature fluctuations in response to the changes in its service environment. In some cases, the temperature gradient could be large, and the resulting thermal stress could be substantial and shall not be ignored. Because temperature fluctuation is a real-life situation, thermal cycling test was developed to understand the thermal fatigue behavior of solder joints.

In passive component manufacturing, materials with vastly different thermal expansion coefficients are utilized in making chip capacitors, resisters, and inductors. Among them are ceramics and glasses. The manufacturing process consists of many steps and is beyond the scope of this chapter. Readers who are interested in this subject should refer to Ref. 61.

In today's passive component manufacturing, the last step is electroplating. Specifically, a matte Sn coating with a thickness of 5–10 µm is plated over a Ni barrier layer. The Ni barrier layer is typically of ca. 5 µm thick, and is also electroplated.

At the end of the manufacturing steps, the passive components could contain many layers of materials (ceramics, glass, Ag, Ni, and Sn). Their relationship is illustrated in Fig. 5.27. These different materials have rather different thermal expansion coefficients. When these components are subjected to temperature cycling, temperature is varied from extremely cold (as low as $-55°$) to extremely warm (as high as $125°C$). Because of the thermal mismatches among these layers of materials, thermal stress develops at various interfaces. After a number of cycles from low to high temperatures, the thermal stress generated can be substantial enough to induce whisker formation. Brusse [62] demonstrated profuse whisker formation of Sn-plated capacitors after thermal cycling the parts for 500 cycles from $-40°C$ to $+90°C$ (Fig. 5.28).

It is important to point out that, until recently, the effect of thermal stress on whisker growth was not recognized by the electronic industry. The awareness was largely brought out by NASA, and companies such as Chippac, Motorola, and the National Electronics Manufacturing Initiative (NEMI) whisker testing task group.

Although whisker growth induced by thermal stress is demonstrated with the chip capacitors, we also find its place in IC fabrications such as leadframe manufacturing. It has been long recognized that the whisker propensity is significantly lower on alloy 42 (a Ni/Fe alloy) than on Cu-based leadframe materials such as alloy 194, alloy 151, and alloy 7025 while employing normal whisker test conditions. These conditions include room and elevated temperature storage with low and high humidity. Recently, it was reported that under temperature cycling conditions (Pascal Oberndorff,

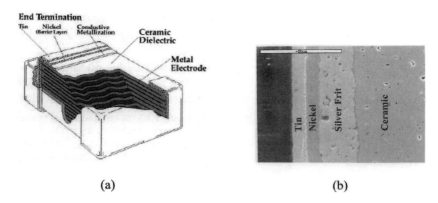

(a)   (b)

Figure 5.27 (a) The cutaway view of a tin-plated multilayer ceramic chip capacitor. (Courtesy of Kemet Electronics Corp. Data Sheet.) (b) Cross-sectional view of the sequence of materials at the metal termination. Tin finish is the final finish. (Photo courtesy of Jay Brusse from NASA.)

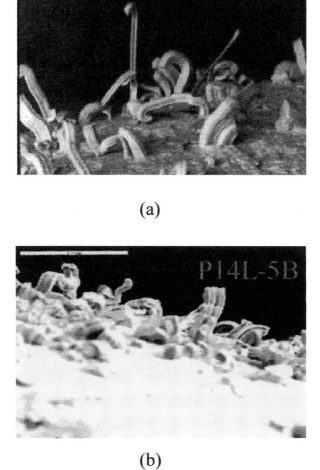

(a)

(b)

Figure 5.28 Examples of the whiskers found on the ceramic chip capacitors after thermal cycling from −40°C to +90°C.

Philips Internal Whisker Investigation, private communications, September 12, 2002), whisker propensity is significantly higher on alloy 42 substrate and whiskers appear quicker than from Cu leadframe materials. It is believed that, in this case, the whisker growth is induced by thermal stress due to CTE mismatch, rather than the irregular intermetallic compound formation such as in the case of the Sn–Cu interface.

The most popular thermal cycling profiles are −55°C to 85°C, and −35°C to 125°C for 500 cycles. Although recently [63], Texas Instruments concluded from their internal studies that a longer cycle time up to 2000 cycles may be needed to ensure confidence of parts reliability.

Overall, there are many new findings on the subject of whisker formation in terms of external mechanical stress and thermal stress. Collectively, we have gained a better understanding on these effects. However, conflicting results still exist. Different and inconsistent whisker storage and test conditions and whisker examination methodology are the main sources of the confusion. Therefore, before any involved discussions can be made on the whisker growth mechanism, it is necessary to examine the whisker investigation methodology.

5.5 WHISKER GROWTH INVESTIGATION METHODOLOGY

5.5.1 Whisker Test Conditions and Observation Methodology

As demonstrated in previous sections, whisker growth phenomenon is very complex. It involves many factors, and they often interrelate to one another. To investigate such a complicated phenomenon, it is critical to establish a systematic and careful sample handling and examination plan. Systematic approach calls for handling samples the same way throughout the lifespan of the whisker study. All samples need to be handled with care to ensure that no physical damages, such as scratch and bending, occur unless they are part of the whisker test matrix. Samples should be stored in an environment where there is no persistent vibration. Temperature and humidity should be well controlled.

In addition to sample handling, frequent and consistent examination of whiskers is equally important. An examination plan entails the followings: (a) examination area: it should be large enough so the results are representative; (b) data points in the observation area: sufficient data points need to be taken so the results have statistical significance; (c) examination method: optical or SEM or both? Samples are best tilted by an angle (45° are often used) for SEM examinations; (d) examination frequency: time zero data are essential. Afterward, twice a month is sufficient.

After whiskers are found, they should be classified into categories for easy data interpretation and analysis. The whisker frequency, length, and diameter are to be estimated. Zhang et al. [64] went a step further. A concept called "whisker index" was introduced to quantitatively characterize and measure the whisker growth propensity by taking into consideration all physical dimensions as well as frequencies of whiskers. An empirical equation was developed to calculate the whisker index [64]. A slightly different version of the whisker index was reported by Whitlaw and Crosby [65] a year later.

Although a systematic sample handling and examination plan is critical in obtaining meaningful results, the most critical part of whisker growth investigation is choosing the right whisker test method. Specifically, it means choosing a set of environmental conditions under which the electroplated Sn parts are stored. Furthermore, it is implied that under such conditions, whiskers will grow most quickly. The determination of this set of conditions requires an understanding of the whisker growth mechanism and the relative importance of the key factors that contribute to whisker formation.

Up to this day, there has been no established standard whisker test method. The fundamental reason is a lack of complete understanding of whisker growth mechanism.

Historically, the following environmental storage conditions have been utilized: room temperature, room temperature/85% relative humidity (RH), 50°C, 50°C/85% RH, and 85°C/85% RH. In addition, thermal bakes at various temperatures such as 125°C, 155°C, and 175°C with different durations have been utilized. Recently, various original equipment manufacturers (OEMs) as well as NEMI's whisker testing group have considered temperature cycling as the accelerating factor. The most common program was to cycle the plated parts between −35°C and 125°C for 500 cycles. Recently, Philips stated that a −55°C to 85°C cycle produces whiskers (Pascal Oberndorff, Philips Internal Whisker Investigation, private communications, September 12, 2002) more readily than the −35°C to 125°C cycle. Although temperature cycling appears to produce whiskers on Sn-plated parts more readily than some other test conditions, questions remain as to whether it is most representative of real-world situations. In addition, a correlation between the whisker test and device reliability needs to be established.

Recently, the E3 group has come up with not only a whisker specification, but also a whisker test method, which has been shown to reproducibly grow whiskers more readily at ambient temperatures than at other accelerated conditions such as elevated temperature, and high-temperature and high-humidity storage (Fig. 5.29).

Their test method calls for using alloy K75 (a Cu alloy, similar to alloy 151 in composition) as substrate; plating 2-μm pure Sn without the use of a Ni barrier layer; storing the plated leadframes at room temperature; and observing whiskers at time zero, first week, second week, etc., up to 8 weeks. If there is no whisker longer than 50 μm during this 8-week period, this Sn finish meets the specification regarding whisker reliability. This test method was derived from extensive internal studies on the part of E3 group. Figures 5.30 and 5.31 explain the basis why they chose K75 as the substrate material and

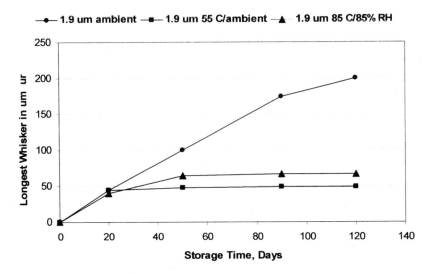

Figure 5.29   Maximum length (in μm) of tin whiskers found on tin finishes stored at different conditions as a function of time. Please note that ambient condition produced the longest whisker. (Data courtesy of Dr. Marc Dittes of Infineon Technologies.)

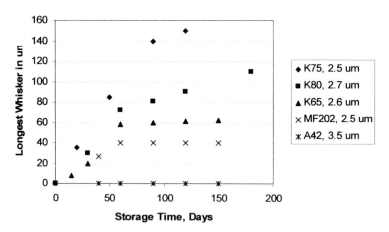

Figure 5.30   Maximum length (in μm) of tin whiskers found on tin finishes on different substrate materials. Please note that the whisker growth rate on K75 appears to be the fastest. (Data courtesy of Dr. Marc Dittes of Infineon Technologies.)

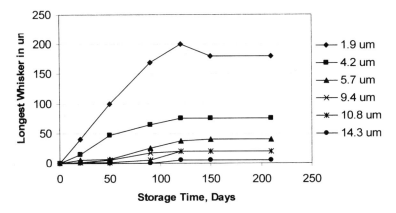

Figure 5.31 Maximum length (in μm) of tin whiskers found on tin finishes plated on K75 substrate with different thickness, varying from about 2 μm to about 15 μm. Please note that the 2-μm finish showed the fastest growth rate. (Data courtesy of Dr. Marc Dittes of Infineon Technologies.)

why 2 μm was the Sn thickness chosen for the whisker test. Figure 5.32 explains why they chose 8 weeks as the test duration. The author believes that this is a more realistic whisker test. It is relatively quick; therefore, it is practical for companies to carry out their own internal whisker tests.

Parallel to finding a standard whisker test method, efforts have been made to characterize quantitatively the stress and microstructure of various Sn deposits. In this regard, some recently developed analytical tools have been utilized to characterize the physical structure and morphology of Sn whiskers. In the following sections, we shall discuss the stress and orientation measurements, and cross-sectional studies of whiskers by FIB.

5.5.2 Stress Measurements

Reports of stress in electrodeposits date back from the middle of the 19th century. Gore reported in 1858 that the inner and outer surfaces of electro-deposited antimony were in a state of unequal cohesive tension. Mills made the first measurement of stress in 1877 by observing the rise or fall of mercury in a thermometer bulb, which was initially chemically coated with Ag. When subsequently plating Cu, Ni, and Fe, a rise in "temperature" (tensile stress) was observed. Plating Cd and Zn resulted in a "temperature" drop (compressive stress). Bouty confirmed Mills's results and developed a formula for converting the "temperature" rise to stress. Since those early days, there

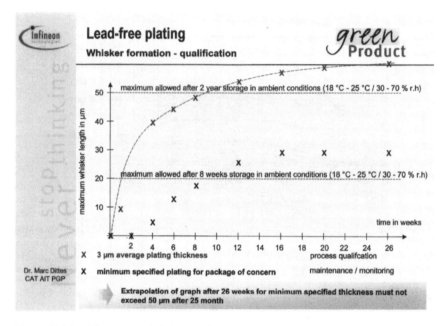

Figure 5.32 Diagram explaining the justifications for (1) whisker test method, (2) whisker specification, and (3) pass or fail criterion.

have been many experiments dealing with the phenomenon of stress in electrodeposits.

Before we talk about stress measurements, it is important that we know what stress we are measuring. The stress mentioned here specifically refers to the deposit residue stress, which is defined as the stress developed within the Sn coating. It can be further classified as intrinsic and extrinsic. Intrinsic stresses are said to develop independently of the substrate, whereas the extrinsic type is due to the interaction of the substrate and the deposit.

There are two ways of measuring stress in electroplated coatings: mechanical and x-ray diffraction.

As its name infers, mechanical method means that the stress is measured mechanically. The simplest and still most widely used mechanical means of measuring deposit stress is the flexible cathode. It consists of a metal strip, insulated on one side, which deflects on being coated on the other side with a stressed deposit. There are a number of different versions of this stress-measuring device that differ mainly in the way deflection is measured. An excellent treatment of mechanical stress methods can be found in the AES Research Project 32 review of Weil [66] in 1970. The advantage of the mechanical methods lies in the fact that stress can be measured in situ (i.e., when electrodeposition takes place). This would allow a quick and direct assessment of deposit stress resulting from a particular chemistry. The stress measured mechanically is the average deposit stress or macrostress.

It is important to recognize that although it is normally not an issue to measure stress in electrodeposits such as Ni, Zn, Rh, etc., it remains a challenge to measure stress in Sn coating because the stress intrinsically is very low for Sn finishes. However, recently, there have been some significant improvements in the mechanical sensor system with the deflection method or flexible cathode method; accurate results have been obtained with Sn deposits [18]. Figure 5.33 demonstrates the hardware of such a device.

X-ray diffraction is widely used as a nondestructive method for stress measurement. This method is based on the changes in spacing between crystal planes induced by stress. If the spacing of the planes is known in the unstressed condition, the stress can be readily calculated. However, because of the codeposition of foreign materials, it is incorrect to assume that the interplanar spacings of an electrodeposited metal are the same as the published values for wrought or cast material. The "two-angle" method does not require knowledge about the spacings between planes in the unstressed condition. By determining the interplanar spacing perpendicular to the surface and at an inclined angle ψ, the unknown quantity (spacing in the unstressed condition) can be mathematically eliminated. The stresses in the two directions are assumed to be related by the conventional elasticity equations, which assume isotropic elastic constants. The following equation is used to calculate stress from the Bragg angles:

$$\sigma = E\cot\theta \, (2\theta_\perp - 2\theta_\psi)/2(1 + v)\sin^2\psi \tag{5.2}$$

where

E: modulus of elasticity of substrate
σ: average deposit stress
θ: Bragg angle
$\theta_\perp$: Bragg angle when specimen surface normal bisects angle between incident and diffracted beams
θ_ψ: Bragg angle when specimen surface normal is inclined by angle ψ
v: Poisson's ratio

It can be seen that the above equation involves a small difference between the two angles $(2\theta_\perp - 2\theta_\psi)$. An angular difference on the order of $0.01°$ corresponds to a stress of 1 kg/mm^2 or 10 MPa. It is difficult to determine the two Bragg angles with sufficient accuracy because of the grain size and the anisotropic nature of the deposit. In addition, if the deposit is textured or highly oriented, the intensity of some diffraction lines may become so weak

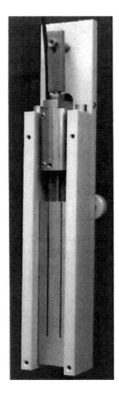 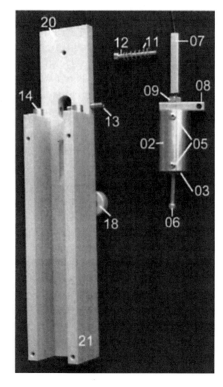

Figure 5.33 Pictures illustrate the mechanics of the hardware to measure stress in the tin layer during deposition.

as the specimen is rotated to the inclined angle that it becomes almost impossible to make sufficiently precise determination of the Bragg angle. That is why we expect relatively large experimental errors in determining the absolute values of the stress for Sn finishes.

Having said that, there are things one can do to improve the accuracy as well as precision of the measurements. First of all, instead of using the standard Cu Kα radiation, one can use either a Co or a Cr radiation source. They are higher energy beams and give deeper penetration depths. Secondly, once can utilize a solid-state two-dimensional detector, which significantly reduces the dark current and increases the signal-to-noise ratio.

In addition, during a specific experiment, one should use a diffraction line at the highest possible angle to improve the precision of the measurements. Furthermore, in the sin$^2 \psi$ test, it is desirable to include as many ψ angles as possible, and in both $-$ and $+$ directions.

With sophisticated state-of-art x-ray diffractometer available today, it has been shown that stress in Sn electrodeposits could be measured with much improved precision utilizing the two-angle method [67]. However, care has to be taken to calibrate the equipment with samples with known stresses, and with both high and low stress standards. For measuring stress in Sn finishes, calibrating the equipment with a stress-free sample is strongly recommended.

It is important to point out that the x-ray method measures not only the microstress but also the microstress. The latter involves using x-ray synchrotron source, which is capable of delivering white x-ray beam, focused to 1 μm. The size of the beam is smaller than the grain size of the deposit so that single grain white beam Laue pattern can be obtained for each individual grain. In such an experiment, Tu revealed that the growth direction of the whisker is ⟨001⟩ whereas the Sn grains in the deposits have a strong texture of ⟨321⟩. In addition, in the same investigation, they found that the overall deposit stress was compressive when whiskers were observed, but the stress level just below the whisker was slightly less compressive than areas that were farther away from the whisker. This is because the stress near the whisker is somewhat relieved by the whisker growth.

5.5.3 Whisker Cross Sections by Focused Ion Beam (FIB) Technology

Focused ion beam technology is an analytical tool developed in the last decade by the semiconductor industry. It is used primarily to obtain cross sections of microvias embedded in semiconductor wafers. It has two unique features. First of all, an FIB resembles a SEM. However, instead of using electron beams such as in the case of SEM, gallium ions generated in a liquid–metal ion source are employed as the imaging beam. In the optical column, these ions are electrostatically focused into a fine beam that scans the specimen surface, releasing secondary electrons for fine structural imaging. Secondly, the high-energy (10–25 keV) Ga beam is like a milling machine. It precisely cuts small trenches in fine circuitries with nanometer accuracy. By simply reducing the intensity of the ion beam, one can effectively "polish" or smooth out the cross sections so that detailed microstructural information such as individual grains and grain boundaries can be revealed. Figure 5.34 gives examples of FIB cross sections of whiskers from (a) a matte Sn finish, (b) a satin bright finish, and (c) a bright finish.

From these photos, it is clear that whiskers grow outward from electroplated Sn coatings. They are "pushed" out of the Sn layer by the need to relieve the compressive stress in the Sn layer.

It is also shown from these photos that the IMC formation at ambient temperature is rather irregular, and its growth is faster at grain boundaries

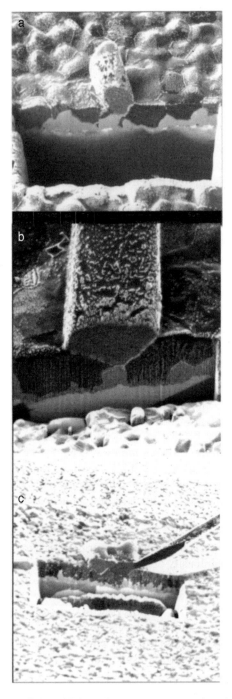

Figure 5.34 Images of the whiskers that were cross sectioned by FIB. Keys: (a) matte finish, (b) satin bright finish, and (c) bright finish. Tin thickness was 3 μm. All samples were stored at room temperature for 18 months before the FIB experiments.

than through the bulk Sn layer. This difference in growth rate generates compressive stress at or near Sn grain boundaries. In turn, a whisker is "pushed" out at the weakest grain boundary when the compressive stress reaches the threshold.

Both the matte and satin bright Sn deposits were 3 μm in thickness, and they were aged at room temperature for 18 months. The FIB photos suggest that the IMC grows faster in a matte finish than in a satin bright finish. Furthermore, the average grain size for matte Sn is about 2 μm, and for satin bright is about 5 μm. This result proves that IMC growth is faster with a smaller grained deposit.

In addition, as it is clearly shown in Fig. 5.34a and b, whiskers are "sitting" on top of the IMC compound. This explains why, in these cases, whiskers contain significant amounts of Cu.

It is worth mentioning that FIB technology was also utilized to understand whisker growth phenomenon with Sn–Cu finishes [68]. Similar observations were made with regard to the driving force for whisker formation.

5.6 WHISKER GROWTH MECHANISM

In seeking the whisker growth mechanism, we are setting out to find out the steps in which the whisker growth takes place, to find out the automatic and consistent response of the whisker growth to various stimuli (factors), and to define and determine the *fundamental* physical, materials, and mechanical principles that govern the whisker growth.

First, let us examine the steps of whisker development. Regardless of the type of Sn finish utilized and the type of whiskers formed, the process of whisker development consists of three steps: (a) incubation or nucleation period; (b) period of growth; and (c) reduction and termination of growth.

The incubation period varies greatly from one case to another. What is the determining factor that defines the incubation time? The author believes that it is a combination of the thermodynamic driving force and the growth kinetics. Specifically, compressive stress is the driving force for whisker formation. If an Sn deposit is tensile-stressed throughout its life, one should not expect to observe whiskers because it is not possible thermodynamically. A threshold of the compressive stress has to be reached before whisker growth becomes thermodynamically favorable.

The incubation time and nucleation period is determined by the time that it takes for the compressive stress to reach the threshold, and the kinetics for whisker nucleation and growth.

What factors govern the kinetics of whisker nucleation and growth? Is it grain size, grain shape, preferred orientation, and/or crystal defects such as dislocations?

As expected, electroplated pure Sn coatings are far from perfect. The number of dislocations in an as-plated Sn layer is likely to be high. The presence of a high concentration of dislocations within the Sn layer presents an unstable microstructure with local sites of high energy content [69]. Any additional stress can produce a change or recrystallization at or around dislocation sites; this lowers the total free energy of the plated layer according to the laws of thermodynamics.

It is most likely that whisker grows as a result of recrystallizations of Sn grains at dislocation sites. Therefore, to reduce whisker growth propensity, one needs to reduce the number of dislocations or to increase the activation energy of recrystallization. When Sn grains have very strong texture, it is usually true that there are fewer dislocation sites. In addition, we know from experience that it is harder for large-grained deposits to recrystallize than small-grained deposits. Therefore, we can deduce that a large, well-polygonized, highly textured Sn deposit has fewer defects such as dislocations, and it is beneficial in whisker reduction when everything else is equal.

All the work reported to date supports above statement. However, the question now becomes, "Is there a specific texture that is beneficial? "How large a grain is large?" Examining finishes that are known to be whisker-resistant, Schetty [46] found that these finishes all have a common orientation, $\langle 220 \rangle$. In a separate and unrelated study, Zhang and Abys [45] observed the minimum whisker formation from a Sn finish that showed only the $\langle 220 \rangle$ and $\langle 440 \rangle$ diffraction lines in the 2θscan. This Sn finish also displayed large, well-polygonized grains. The average grain size determined by TEM was about 5 μm. Incidentally, in his work, "A Laboratory Study of Sn Whisker Growth," Dunn observed that the recrystallization rate of Sn was dependent on the crystallographical orientation of the grains and the growth direction. Specifically, he observed that whisker growth directions were of $\langle 101 \rangle$, $\langle 100 \rangle$, and $\langle 111 \rangle$ orientation. Please note that $\langle 110 \rangle$ was missing in his and other studies cited previously.

Once the whisker nucleation begins, the growth period most likely is determined by the sustained compressive stress in the Sn layer, and the availability of the Sn atoms that diffuse toward the nucleation site. When Sn grain size is small, there are a lot of grain boundaries; therefore, there are many paths for Sn atoms to diffuse to the nucleation site.

Once the whisker forms, the compressive stress in its vicinity is reduced or partially released. This point is demonstrated by Choi et al. [41], where the researchers utilized the synchrotron x-ray source to measure microstress within the Sn layer. They were able to measure the stress on the whisker and next to the whisker. They found that the compressive stress is lower in the vicinity of the growing whisker. It is important to point out that Choi's work also demonstrates that there is a threshold of the compressive stress. When the stress level is higher than this value, whisker growth becomes inevitable. However, once the compressive stress is reduced to below this level, no additional whiskers grow until additional compressive stress builds up again in the Sn layer.

As one can see, the above description of whisker growth mechanism is still rather preliminary. To obtain a complete understanding of the whisker growth mechanism, we need to bear in mind that the compressive stress can be generated chemically (e.g., IMC), thermally (e.g., CTE mismatch), and mechanically (e.g., bending and scratching). Therefore, we should focus on understanding better these origins of the compressive stress. We should also focus on better understanding where the whisker nucleation site is and how this nucleation site is influenced by pretreatment, plating chemistries and process parameters, deposit properties, as well as environmental stimuli after plating.

Independent of the compressive stress theory, there is so-called "grain-boundary-free energy theory"[73]. The essence of this theory is that the grain-boundary-free energy is sufficient for spontaneous recrystalization of the tin layer. Although it also recognizes the importance of recrystalization as the main mechanism for whisker growth, it believes that the grain-boundary-free energy alone is the driving force. In their grain-boundary-free energy calculation, Tsuji and his co-workers showed that the grain-boundary-free energy per unit volume for a matte tin deposit is inversely proportional to the square root of the deposit thickness, whereas it does not change for a bright tin deposit, which is $0.6 \times 10^6 J/M^3$. For a matte tin deposit, it drops steeply from $0.25 \times 10^6 J/M^3$ at 0.5 μm thickness to $5 \times 10^4 J/M^3$ at 10 μm thickness. They argued that the activation energy for recyrstalizing tin requires 10^4 to $0.6 \times 10^6 J/M^3$. Since the grain-boundary-free energy per unit volume of a bright tin finish is about $0.6 \times 10^6 J/M^3$, it explains why bright finish is very prone to whisker formation. On the other hand, a 10 μm matte tin finish has the grain-boundary-free energy of $5 \times 10^4 J/M^3$, at the lower end of the required energy. That is why it has a much lower propensity for whisker growth. It also explains qualitatively why a thin, matte finish has a much higher propensity for whisker growth than a thicker one.

Though both theories explain many of the experimental results, neither explains all the results. Quantitative experimental as well as modeling wotk is critical and needed to fully understand the whisker growth mechanism.

5.7 WHISKER PREVENTION

To prevent whisker formation, we need to develop and maintain a tensile stress in the Sn deposit. Therefore, it is beneficial to have plating chemistries that could yield tensile-stressed Sn deposit.

In addition, we could use a barrier layer such as Ni to stop or reduce the intermetallic compound formation at the Sn–substrate interface; we could anneal the parts at temperatures above 100 °C or reflow the parts to relieve the residue stress; we could rethink and redesign the materials we use in making chip capacitors, transistors, and inductors to reduce the CTE mismatch; and we could be more stringent with our process control parameters during substrate pretreatment and Sn plating. With these precautions, we should be able to significantly minimize the whisker growth propensity.

5.8 WHISKER APPLICATIONS

Although whiskers can have detrimental effect in electronic device by shorting the circuits, they sometimes bring benefits in our daily lives. The author would like to provide a few interesting examples.

In a recent documentary by BBC News On-line [70], scientists have discovered that a seal uses its whiskers to detect the watery trail of a swimming fish. The work was carried out by a German team, which used a miniature submarine to simulate the trail made by a swimming fish. Two captive seals were able to locate the vessel in the dark.

"We found that a seal was able to follow a hydrodynamic trail using its whiskers," coresearcher Bjorn Mauck of the University of Bonn, Germany, told BBC News On-line. "This is the first time that an animal has been shown to do this." It is interesting to note that when the seal's whiskers were covered with stocking, he failed to locate the submarine.

In the technology world, it is not difficult to find unique applications of various kinds of whiskers. For instance, SiC whisker-reinforced ceramic composites were innovations that came into prominence for potential structural applications because of the significant improvements in the mechanical properties these materials offered as compared to the monolithic materials. The incorporation of SiC whiskers into alumina ceramics resulted in increases in strength, fracture toughness, thermal conductivity, thermal shock resistance, and high-temperature creep resistance.

SiC whiskers used for reinforcement are discontinuous, rod-shaped or needle-shaped fibers in the size range of 0.1–1 μm in diameter and 5–100 μm in length. Because they are nearly perfect single crystals, the whiskers typically have very high tensile strengths (up to 7 GPa) and elastic modulus (up to 550 GPa). SiC whisker-reinforced ceramics was introduced commercially for cutting tool applications in 1985 by Greenleaf Corporation. In one reported case history, changing from a conventional tool to a SiC whisker alumina one reduced a 5-h machining operation of inconel to 20 min [71].

In another application [72] by Containerless Research Inc., precision, single-crystal microwhiskers have been grown epitaxially from ⟨111⟩ silicon wafers (Fig. 5.35) or $\langle 111 \rangle$ silicon posts (Fig. 5.36) with finely controlled microscopical dimensions. Whisker-covered areas of 1, 2, 3, or 5 mm diameter on a 10-mm^2 substrate with an array spacing of 15, 30, and 60 μm can be precisely controlled (Fig. 5.37). The whisker diameters are 3, 4, and 5 μm, and are 30–100 μm tall. In addition, the whisker tips can be sharpened to radii less than 10 nm, which make them ideal for scanning tunneling microscopy tips.

Nanometallic whiskers have been produced by ion track technology developed at GSI Materials Research [73]. Scientists were able to produce stochastic arrays of monocrystalline metal whiskers at nanoscale for magnetic sensors, flat screens, and trace impurity sensors applications. Fig. 5.38 illustrates a free standing array of the metal whiskers.

As one can surmise, the most important property employed in the above applications regarding whiskers is their mechanical strength and small dimensions. Needless to say, there are many other applications of whiskers in the microtechnology and nanotechnology world that are not included in this chapter.

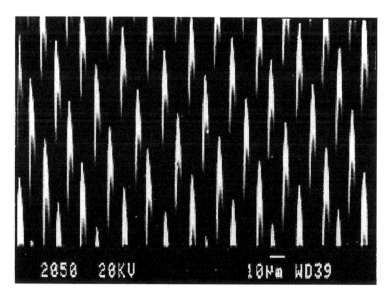

Figure 5.35 Whiskers grown epitaxially from $\langle 111 \rangle$ silicon wafers with finely controlled microscopical dimensions. The whiskers are fabricated in precise arrays with uniform diameter and length by the vapor–liquid–solid (VLS) process. (Photo courtesy of Containerless Research Inc.)

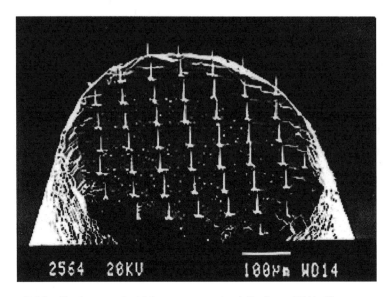

Figure 5.36 Single-crystal whiskers grown epitaxially from ⟨111⟩ silicon posts with finely controlled microscopical dimensions. The whiskers are fabricated in precise arrays with uniform diameter and length by the vapor–liquid–solid process. (Photo courtesy of Containerless Research Inc.)

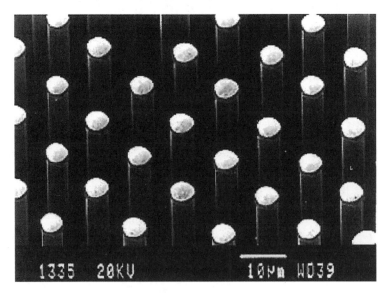

Figure 5.37 Same as in Fig. 5.36. The whisker tips were gold-coated. The array spacing is ca. 15 μm. (Photo courtesy of Containerless Research Inc.)

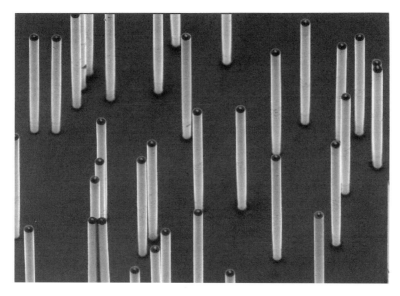

Figure 5.38 Free standing metallic whiskers form GSI Materials Research. (Photo courtesy of GSI Materials and Research.)

5.9 CONCLUSION

Whisker growth phenomenon is a very complex subject. To understand this phenomenon, it requires understanding of chemistry, electrochemistry, metallurgy, crystallography, and mechanical engineering. In addition, it requires systematic and precise work, which is usually tedious and time-consuming.

Unfortunately, the move toward becoming lead-free does not give large electronic companies the luxury of waiting for the research and understanding to be completed. Instead, they have to take a proactive approach and make the best of the knowledge available today.

Even though users of electrodeposited Sn wish to have guarantees from the chemical suppliers to provide "whisker-free" Sn finishes, it simply cannot be done. As it is demonstrated throughout this chapter, there are many factors independent of the plating chemistry that contribute to whisker growth. As the author pointed out previously, there are certain areas that critically need more in-depth and systematic investigations. Nobody can do this alone. The best approach would be for the chemical suppliers to work with their customers further up in the supplier chain to look closely at each step of the manufacturing processes, and to optimize the conditions so that lowest whisker propensity is achieved at the end.

ACKNOWLEDGMENTS

The author especially would like to thank Joe Abys, Chen Xu, Chonglun Fan, and Anna Lifton for their contributions to the whisker research, and Oscar Khaselev and Igor Zavarine for their contributions to the spectro-electrochemical studies conducted at EC&S. The author would also like to thank organizations such as NEMI, NIST, and IPC for their support and encouragement. The author is deeply indebted to many friends, colleagues, and customers in the electronics industry whom she has had the privilege of knowing and working with. Without them, this chapter would have not been possible.

REFERENCES

1. HY Hunsicker, LW Kenspf. Growth of whiskers on Sn–aluminum bearings. Q Trans SAE 1:6, 1947.
2. C Herring, JK Galt. Elastic and plastic properties of very small metal specimens. Phys Rev March 85:1060–1061, 1952.
3. Food and Drug Administration. ITG #42: tin whiskers—problems, causes and solutions (http://www.fda.gove/inspect_ref/itg/itg42.htm), accessed March 1, 1986.
4. K Heutel, R Vetter. Problem notification: Sn whisker growth in electronic assemblies. Internal memo, February 19, 1988.
5. B Nordell. Air Force links radar problems to growth of Sn whiskers. Aviat Week Space Technol June 20, 1986, 65–70.
6. J Richardson, B Lasley. Sn whisker initiated vacuum metal arcing in spacecraft electronics. Proceedings of the 1992 Government Microcircuit Application Conference. Vol. XVIII. November 10–12, 1992, pp 119–122.
7. M Ohring. Reliability and Failure of Electronic Materials and Devices. San Diego: Academic Press, 1998.
8. DT Hawkins. Metal whiskers 1945–1975, Bell Labs internal document, a comprehensive collection of 886 references on metal whiskers.
9. SM Arnold. Repressing the growth of Sn whiskers. Plating 1966; 53:96–99.
10. Y Zhang, G Breck, F Humiec, K Murski, JA Abys. An alternative surface finish for Sn–lead solders: pure tin. Proceedings of the SMI'96, San Jose, CA, 1996, pp 641–649.
11. Y Zhang. U.S. Patent 5,750,017.
12. Y Zhang. Can electroplated pure Sn be utilized as a final surface finish for PWBs? CircuiTree November 1999.
13. S Winkler, B Hom. A look a past reveals a lead-free drop-in replacement. High-Density Interconnect (on-line) March 2001.
14. R Schetty. Minimization of Sn whisker formation for lead-free electronics finishing. Proceedings of the IPC Works Conference, Miami, FL, September 2000.

15. KM Cunningham, MP Donahue. Sn whiskers: mechanism of growth and prevention. 4th International SAMPE Electronics Conference, 1990, p 569.
16. M Jordan. Lead-free Sn alloys as substitutes for Sn–lead alloy plating. Trans IMF 75(4):149–153, 1997.
17. Thirty-seven new R&D projects selected for 1999 Advanced Technology Program Award (10/7/99). (http://www.atp.nist/atp/archval.htm). Technologies of the next century. Ind Phys December 1999, pp 14–15.
18. Y Zhang, C Xu, J Abys, C Fan, A Vysotskaya. Understanding whisker phenomenon—whisker index and tin/Cu, tin/nickel interface. Proceedings of the APEX Expo, San Diego, CA, 2002.
19. K Fujiwara, R Kawanaka. Observation of the Sn whisker by micro-Auger electron spectroscopy. J Appl Phys 51(2):6231–6232, 1980.
20. NEMI Whisker Test Group Definition, July 2001.
21. ME Williams, CE Johnson, KW Moon, GR Stanford, CA Handwerker, WJ Boettinger. Whisker formation on electroplated Sn–Cu. Proceedings of Sur-Fin'2002, Navy Pier, Chicago, IL, 2002, pp 19–30.
22. BD Dunn. A laboratory study of Sn whisker growth. Report STR-223. Noordwijk, The Netherlands: European Space Agency (54 pp.).
23. JF Smith. Whisker growth on electrodeposited tin. Trans Inst Met Finish 45 (917):9–11, 1967.
24. Y Zhang, C Xu, C Fan, J Abys. Electroplated Sn and the whisker phenomenon. Proceedings of UIM Conference, UIM, Germany, May 2000.
25. P Harris. The Growth of Sn Whiskers. International Sn Research Institute, 1994 (publication no. 734).
26. GW Stupian. Sn whiskers in electronic circuits. Aerospace Technical Report No. 92, 1992, pp 2925–2917.
27. SC Britton. Spontaneous growth of whisker on Sn coatings: 20 years of observation. Trans Inst Met Finish 1974; 52:95–102.
28. Product/Process Problem Alert Bulletin 9202. The Aerospace Corporation, May 14, 1992.
29. JH Davis, MJ Skove, EP Stillwell. Superconducting transition temperature and resistivity of Sn whiskers as a function of uniaxial tension. Solid State Commun 4(11):597–600, 1966.
30. WW Webb. Stage steps in the resistive transitions of superconducting microcrystals. Sci Technol Aerosp Rep 7(3):557, 1969.
31. TE Lukens, RJ Warburton, WW Webb. Onset of quantized thermal fluctuations in "one-dimensional" superconductors. Phys Rev Lett 25(17):1180–1184, 1970.
32. NAJ Sabbagh, HJ McQueen. Sn whiskers: causes and remedies. Met Finish March 1975.
33. AJ Bevolo, JD Verhoeven, M Noack. A Leels and Auger study of the oxidation of liquid and solid tin. Surf Sci 134:499, 1983.
34. RA Konetzki, YA Chang, VC Marcotte. Oxidation Kinetics of Pb-Sn alloys. J Mater Res 4:1421, 1989.
35. RG Miller, CQ Bowles. Oxidation kinetics of Pb-Sn alloys. Oxid Met 33:95, 1990.

36. SC Britton, K Bright. Metallurgia 56:163, 1957.
37. KN Tu. Irreversible processes of spontaneous growth mechanism of Sn whiskers. Phys Rev B 49(3):2030, 1994.
38. I Zavarine, O Khaselev, XP Gao, Y Zhang. Application of spectroelectrochemistry in electroplating industry. Proceedings of SurFin'2001, Nashville, TN, June 2001.
39. Y Zhang. ATP Project Lucent9H3021. Technical Quarterly Report, January 2000–December 2002.
40. Y Zhang, C Xu, C Fan, JA Abys. Whisker growth and prevention. J Surf Mount Technol 13(4):1–9, 2000.
41. WJ Choi, TY Lee, KN Tu, N Tamura, RS Celestre, AA MacDowell, YY Bong, L Nguyen, GTT Sheng. Structure and kinetics of Sn whisker growth on Pb-free solder finish. Fifty-Two Electronic Component and Technology Conference Proceedings, San Diego, CA, 2002, pp 628–633 (IEEE catalog no. 02CH3734-5).
42. BD Dunn. ESA Sci Tech Rev 2:1–10, 1976.
43. SM Arnold. Growth of metal whiskers on electrical components. Electr Manuf 110–114, November 1954.
44. SM Arnold. Repressing the growth of tin whiskers. Plating 53:96–99, 1966.
45. Y Zhang, JA Abys. A unique electroplating Sn chemistry. Circuit World 25(1): 35, 1998.
46. R Schetty. Sn whisker growth and the metallurgical properties of electrodeposited Sn. Proceedings of the JPC/JEDEC International Conference on Pb-Free Electronic Assemblies, San Jose, CA, 2002, pp 92–94.
47. NEMI Whisker Test Group Communication, October 2002.
48. KN Tu. Interdiffusion and reaction in bimetallic Cu–Sn thin films. Acta Metall 21(94):347, 1973.
49. Klein-Wassink. Soldering in Electronics. Ayr, Scotland: Electrochemical Publications, Ltd., 1989.
50. BG LeFevre, RA Barczykowski. Wire J Int 18:66, 1985.
51. CA Mackay. Welding Met Fab January/February, 1979, p 53.
52. V Simic, Z Marinkovic. Room temperature interactions in copper-metal thin film couples. J Less-Common Met 72:133, 1980.
53. HAH Steen, A Bengston. Brazing Solder 13:28, 1987.
54. DA Unsworth, CA Mackay. A preliminary report on growth of compound layers on various metal bases plated with tin and its alloys. Trans Inst Met Finish 51:85, 1973.
55. BZ Lee, DN Lee. Spontaneous growth mechanism of Sn whiskers. Acta Mater 46(10):3703–3714, 1998.
56. KN Tu. Cu/Sn interfacial reactions: Sn film case versus bulk case. Mater Chem Phys 46:217–223, 1994.
57. Y Zhang. Understanding whisker growth phenomenon. Proceedings of CALCE Meeting, University of Maryland, October 10, 2002.
58. A Ugli. Proceedings of IPC Annual Meeting, New Orleans, TN, November 2002.
59. Y Zhang, R Schetty. Whisker growth: the substrate effect and beyond. Proceedings of IPC/JEDEC Meeting, Taipei, Taiwan, December 2002.

60. AC Tan. Sn and Solder Plating. A Technical Guide. London: Chapman and Hall, 1992.
61. RJK Wassink. Soldering in Electronics. Second Ed. England: Electrochemical Publications Limited, 1989, Ch. 8.2.
62. J Brusse. Sn whisker observations on pure Sn-plated ceramic chip capacitors. Proceedings of Surfin'2002, Chicago, IL, June 2002.
63. Texas Instruments. NEMI whisker testing group meeting minutes, September 12, 2002.
64. Y Zhang, C Xu, C Fan, A Vysotskaya, J Abys. Understanding whisker phenomenon: Part I. Growth rate. Proceedings of SurFin'2001, Nashville, TN, June 2001.
65. K Whitlaw, J Crosby. An empirical study into whisker growth of Sn and Sn ally electrodeposits. Proceedings of SurFin'2002, Chicago, IL, June 2002.
66. R Weil. The origins of stress measurements. AES research project 32. Plating December 1970, p 1231.
67. G Gore. On the properties of electrodeposited antimony. Trans Roy Soc (London), Part I:158, 1858.
68. EJ Mills. On electrostriction. Proc Roy Soc, 26:504, 1877
69. M Bouty. Pressures caused by electrodeposits. Compt Rend 88:714, 1879.
70. C Xu, Y Zhang, C Fan, J Abys. Understanding whisker phenomenon: Part II. Competitive mechanisms. Proceedings of the APEX Expo, San Diego, January 2002.
71. GTT Sheng, CF Hu, WJ Choi, KN Tu, YY Bong, L Ngugen. Sn whiskers studies by focused ion beam imaging and transmission electron microscopy. J Appl Phys 92:64–69, 2002.
72. DR Gabe. Principles of Metal Surface Treatment and Protection. 2d ed. Oxford: Pergamon Press, 1978.
73. K Tsuji. Role of grain-boundary-free energy and surface-free energy for tin whisker growth. Proceedings of SurFin' 2003, Midwest Airlines Center, Milwaukee, Wisconsin, 2003, pp 169–186.
74. Seals use whiskers to track prey. BBC News On-line, July 6, 2001, Friday.
75. SiC whisker-reinforced ceramic composites (http://www.ornl.gov/MC-CPS/sic.htm), accessed December 4, 2002.
76. Precision, Single Crystal Silicon Micro-Whiskers. Containerless Research Inc. (http://www.containerless.com/whisker.htm), accessed December 4, 2002.
77. Nano Whiskers. GSI Materials Research (http://www-wnt.gsi.de/mr/nano_whiskers.htm), accessed December 4, 2002.

6

Mechanical Evaluation in Electronics

William J. Plumbridge
The Open University, Milton Keynes, United Kingdom

6.1 INTRODUCTION

Electronics is undoubtedly the largest and most rapidly changing industrial sector. It is facing the twin challenges of miniaturization and environment. The first has led to a much greater focus upon mechanical behavior and structural integrity of interconnections, particularly since the emergence of Surface Mount Technology (SMT). The second has necessitated the development of a new class of solder alloy that contains no lead. Electronics equipment is generally conveniently small and relatively inexpensive, and this has facilitated whole scale testing, usually under accelerated service-like conditions, prior to marketing. The number of parameters associated with the printed wiring board (PWB)—component type and arrangement, and service conditions—is substantial. The common practice of extensive empirical testing is becoming unwieldy, inefficient, and sometimes incapable of meeting reliability requirements. There is much potential in adopting the strategies of other high temperature engineering applications, such as power generation, and attempting to design interconnections from first principles. Such an approach is sometimes described in the electronics field as the "Physics of Failure." This is not to suggest that empirical testing should be replaced, but rather that a better balance between evaluation modes and a greater mutual appreciation should be sought.

To this end, the present chapter examines the different means of mechanical evaluation of electronics equipment performance. In view of the comments above, emphasis is placed upon obtaining an understanding of the mechanisms responsible for failure and the need to obtain information on mechanical behavior under conditions that are appropriate to those experienced in service. Without such data, computer-based modeling is a futile exercise. Electronics equipment may fail in a variety of ways—many of which are governed by the mechanical properties of the materials involved. There is therefore need for an awareness of both the failure mechanisms and the specific mechanical properties which influence them. Because the performance of soldered interconnections may involve a wide range of disciplines and to provide a scientific and engineering foundation, the general characteristics of the materials associated with interconnections and the common mechanically related failure modes are first investigated. The determination of relevant mechanical properties is then discussed and the challenges of comparability of data are considered. Finally, the most common failure mode, thermomechanical fatigue (TMF), is considered in more depth. Some specific examples of mechanical behavior will be cited, for illustrative purposes only, to demonstrate particular points. A more comprehensive treatment of mechanical properties is presented elsewhere in this text.

6.2 MATERIALS AND THEIR MECHANICAL PROPERTIES

Electronics equipment is somewhat unusual as all categories of engineering materials may exist in close proximity. For example, a printed circuit board is often a fiber-reinforced polymer (it is a composite); a chip may be located on a ceramic, and the tracking and solder will be metallic materials. While acknowledging the importance of physical properties, such as electrical and thermal conductivity and thermal expansivity, the general mechanical characteristics of the material classes above are now summarized in order to establish a basis for understanding more complex behavior in service. In addition, examples of constitutive relationships that describe mechanical behavior will be presented to demonstrate the significance of reliable and appropriate mechanical properties in modeling, life prediction, and design.

6.2.1 Materials Response to Monotonic Stressing

Under the action of an increasing stress, metals usually exhibit elasticity, plasticity, and a maximum in stress is followed by necking and fracture. The slope of the linear elastic portion of the stress vs. strain plot is the modulus, and the stress at termination of elastic behavior is the yield stress. (The yield stress of solders is quite small.) The extent of deformation prior to fracture is known as ductility. Ceramics display only elastic behavior until fracture, which is associated with cracking and very limited deformation (brittleness). Polymers may exhibit both characteristics above according to the temperature. Above the glass transition temperature, T_g, extensive deformation due to mechanisms quite unlike those in metals may follow a small degree of elasticity. Below this temperature, polymers exhibit ceramic-like behavior. In all material categories, the maximum stress attained is the tensile/compressive/shear strength according to the mode of stressing employed. Composites are physical mixtures and exhibit the average properties of their components, taking into account the proportions of each. When the reinforcing agent is asymmetric, such as fibers or wires, the composite becomes anisotropic and has different properties in different orientations. For example, in a common FR-4 (epoxy resin-impregnated fiberglass cloth) PWB, typical values for the coefficients of thermal expansion in the longitudinal and transverse directions are 12–18 ppm per °C, whereas in the thickness direction they are 5–6 times greater [1]. These characteristic features of monotonic behavior are summarized in Fig. 6.1. The fracture strains of brittle materials and the yield strains of metals are generally less than 1% (the yield strain of solders is around 0.1–0.2%). The amount of deformation prior to the attainment of maximum strength is between about 3% and 7% for common solder alloys.

Electronic equipment is required to operate over a range of temperatures. As a general rule, with increasing temperature, the strength of metals falls; that of ceramics is largely unaffected until very high values are reached; polymers also lose strength, although brittle polymers may become ductile if

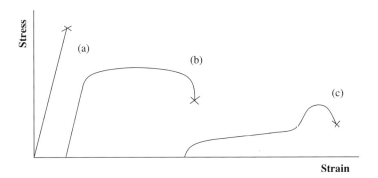

Figure 6.1 Schematic overview of mechanical behavior of different categories of material: (a) ceramics, polymers below their glass transition temperature T_g, non-ductile metals; (b) ductile metals; and (c) polymers above T_g.

the glass transition temperature is exceeded. The rapid application of stress produces high strain rates, which tends to restrict ductility. Very high strain rates are associated with impact damage. A useful property for assessing the strength of nonductile materials that contain cracks, or other defects, is the fracture toughness, K_{IC}, defined as

$$K_{IC} = \sigma Y \sqrt{a} \qquad (6.1)$$

Where a is the crack length in a specimen subject to a stress σ. Y is a geometrical correction factor.

To model monotonic behavior is relatively straightforward in the absence of time-dependent effects. The stress–strain relationship for elastic behavior simply involves the appropriate modulus, E or G, for tensile or shear conditions.

$$\text{Strain} = \sigma/E \text{ or } \tau/G \qquad (6.2)$$

Where σ and τ are the tensile and shear stresses, and ε and γ are the corresponding strains. Subsequent plastic behavior may be modeled as either perfectly plastic, linear work hardening, or in a more sophisticated manner to depict the "true" curved shape of the plastic line (Fig. 6.2). The complete constitutive equation for the linear work hardening case becomes

$$\sigma = \sigma_y + m\left(\varepsilon - \frac{\sigma_y}{E}\right) \qquad (6.3)$$

$$\varepsilon = \frac{\sigma_y}{E} + \left(\frac{\sigma - \sigma_y}{m}\right) \qquad (6.4)$$

Where m is the slope of the plastic line and is known as the work hardening coefficient. All the terms are sensitive to temperature and strain rate, which

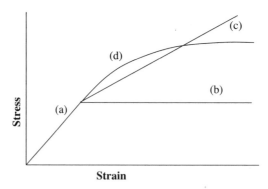

Figure 6.2 Constitutive stress–strain relationships in tension: (a) elastic behavior; (b) perfectly plastic; (c) linear work hardening; and (d) variable work hardening.

provides some insight on the complexity of, and the requirements for, reliable modeling.

6.2.2 Time-Dependent Mechanical Behavior

The converse of impact loading is the situation in which a body is loaded to a fixed value and left in that condition for a prolonged period. In some applications, this may involve many years. The material continues to deform under the action of the constant load (or stress), and this process is known as creep. Again, temperature is a critical parameter, and for metals, creep must be considered when values of the homologous temperature, T_h (the ratio of the current temperature to the melting temperature, expressed in degrees Kelvin), exceed about 0.4. For the conventional Sn–37Pb solder, this is equivalent to $-90°C$, which means that the creep behavior should be considered in all applications. The generally higher melting points of the lead-free solder alloys raises this critical temperature to $-75°C$, which does not materially alter the situation. Ceramics and polymers below their glass transition temperature do not exhibit creep at the temperatures of interest in electronics applications.

There are three stages of creep (Fig. 6.3). After initial instantaneous elastic deformation on loading, primary creep is denoted by a continuous fall in creep strain rate. Secondary creep is characterized by a period in which the creep strain rate remains essentially constant, and in tertiary creep the creep strain rate increases continuously until rupture. Secondary creep is sometimes described as steady-state creep, and the value of the creep strain rate during this stage is the minimum for the test and that used in design calculations. For solders, it generally occupies between 20% and 40% of the total lifetime [2].

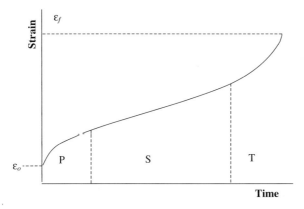

Figure 6.3 The stages of creep (schematic) ε_o and ε_f are instantaneous and final creep strains; P, S, and T are the primary, secondary, and tertiary stages.

The other important terms in creep behavior are the applied stress and the time to rupture, t_r, or alternatively, the time to achieve a specific amount of deformation. Creep ductility indicates the level of deformation prior to fracture—low ductilities being associated with sudden failure at limited strains.

The minimum creep rate may be linked with the applied stress, σ, by a series of equations according to the dominant creep mode.

$$\dot{\varepsilon}_m \propto \sigma^n \text{ (power law creep)} \tag{6.5}$$

$$\dot{\varepsilon}_m \propto \exp (A\sigma) \text{ (exponential creep)} \tag{6.6}$$

$$\dot{\varepsilon}_m \propto [\sin h (B\sigma) \text{ (combination creep)} \tag{6.7}$$

where A and B are temperature-dependent constants.

Creep is highly sensitive to both applied stress level and to test temperature. As a thermally activated process, the creep rates increase exponentially with temperature. The effect of stress is dependent upon the controlling creep mechanism. Values of the stress exponent n may vary between 1 and over 15, which reinforces the need to identify the dominant mechanisms in service and for accuracy in computing stress values (Fig. 6.4). A popular generalized expression for creep is

$$\dot{\varepsilon}_m = A\sigma^n g^{-p} \exp \left(\frac{-Q}{RT} \right) \tag{6.8}$$

where g is the grain size, A, n, and p are constants, and Q is the activation energy of the dominant creep process. Determination of the value of Q enables the identification of the controlling creep mechanism to be made.

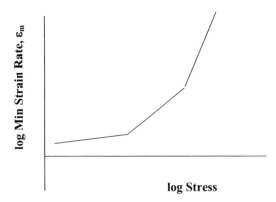

Figure 6.4   Minimum (or steady-state) creep rate vs. applied stress showing transitions in the controlling mechanisms.

Numerous expressions exist for the prediction of creep life although none is universally applicable. Among the most popular are those based upon the Monkman–Grant equation,

$$\dot{\varepsilon}_m t_r = C \tag{6.9}$$

where $\dot{\varepsilon}_m$ is the minimum creep strain rate, t_r is the time to rupture, and C a constant.

Closely related to creep is stress relaxation, which is the time-dependent reduction in stress levels in a body deformed to, and held at, a fixed strain. Stress decays logarithmically as described by:

$$\sigma_t = \sigma_o - A\ln(1 + Bt) \tag{6.10}$$

The processes enabling stress relaxation are important in damage accumulation during dwell periods in the temperature (strain)–time cycle.

6.2.3 The Effects of Repeated Stresses or Strains

Electronics equipment rarely experiences a single constant or a continuously increasing stress. More common are stress or strain fluctuations induced by the service conditions, and these give rise to fatigue failure (failure due to a number of cycles, the magnitude of which is insufficient to cause failure in a single application). The key design parameters are the stress, or strain, amplitude, and the number of cycles, N_f, necessary for failure. The process involves the initiation and gradual growth of cracks until the remaining section of the material can no longer support the applied load. It is by far the most common mode of failure in engineering applications.

The majority of materials display an increase in fatigue endurance as the stress range diminishes. Data in this form are difficult to analyze or to use in expressions for life prediction. Strain is a more fundamental property to materials behavior than stress, because it is a measure of the extent to which the material is deformed. Plots of strain range against numbers of cycles to failure are more useful to the designer, particularly when plotted on logarithmic scales, when linear relationships often exist.

The total strain, ε_t, is made up of an elastic component, ε_e, and a plastic component, ε_p. Similarly, in terms of strain ranges,

$$\Delta\varepsilon_t = \Delta\varepsilon_e + \Delta\varepsilon_p \tag{6.11}$$

A plot of $\log \Delta\varepsilon_t$ against $\log N_f$ is a curve as shown in Fig. 6.5. However, under conditions where either $\Delta\varepsilon_e$ or $\Delta\varepsilon_p$ dominates, plotting these terms against N_f gives a linear relationship. Hence, either

$$\Delta\varepsilon_e N_f^\alpha = C_1 \text{ (Basquin) or} \tag{6.12}$$

$$\Delta\varepsilon_p N_f^\beta = C_2 \text{(Coffin-Manson)} \tag{6.13}$$

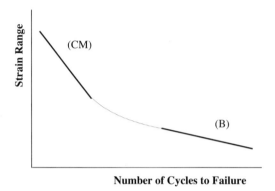

Figure 6.5   Strain range vs. number of cycles to failure showing the domains of elastic (Basquin) and plastic (Coffin–Manson) dominance.

For solders, the extent of elastic strain is small, so that in most cases, $\Delta\varepsilon_\rho \gg \Delta\varepsilon_e$, and a graph of either $\Delta\varepsilon_t$ or $\Delta\varepsilon_\rho$ against number of cycles to failure is generally linear. This is the Coffin–Manson expression that is used as a basis in many life-predictive methods for solder joints.

A hysteresis loop is a convenient means of depicting stress–strain relationships during fatigue (Fig. 6.6). The "width" of the loop is equal to the plastic strain range, $\Delta\varepsilon_\rho$, and the elastic strain range, $\Delta\varepsilon_e$, is given by the difference between this and the total strain range. The area enclosed by the loop is a measure of the plastic strain energy required to produce the observed deformation, and the associated elastic strain energy component is the product of the stress and elastic strain ranges. This type of representation is very convenient when considering more complex and realistic cycles. During strain-controlled fatigue, the stress range necessary to maintain the strain limits may increase (cyclic hardening) or decrease (cyclic softening). These changes may be due to deformation processes or to cracking. A constitutive equation, equivalent to that described for monotonic loading, relates the stress and strain ranges during fatigue, i.e.,

$$\Delta\sigma = k(\Delta\varepsilon_p)^n \tag{6.14}$$

where $\Delta\sigma$ is the stress range at half life, k and n are the cyclic strength coefficient and the cyclic strain hardening exponent, respectively.

An alternative methodology for design, defect tolerance, asks the question "how long (or how many cycles of stress or strain) will it take a pre-existing defect to grow to catastrophic dimensions under service con-

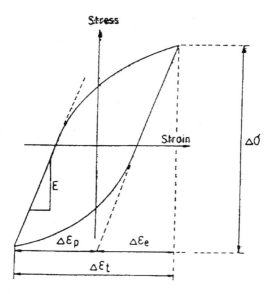

Figure 6.6 Mechanical hysteresis loop showing the complete stress–strain relationship during a fatigue cycle.

ditions?" Under fatigue, the crack growth rate, da/dN, may be correlated with the crack tip stress intensity range ΔK, by

$$\frac{da}{dN} = C(\Delta K)^m \tag{6.15}$$

Where m and C are constants. Integration of this expression between the initial crack size and the critical dimension at failure, determined from fracture toughness values, provides an estimate for the number of cycles to failure, N_f. This approach is based upon linear elastic fracture mechanics (LEFM) which implies limited plasticity. Its application to interconnections is impaired by the ductile nature of solders and the existence of short cracks, the growth of which often does not follow the above expression. Alternatives to stress intensity, such as ΔJ or ΔW (the strain energy density per cycle), may be employed instead and the form of Eq. (6.15) is retained [3,4]. For example, using the hysteresis loop in Fig. 6.6, the total strain energy density, ΔW, is given by

$$\Delta W_t = \Delta W_e + \Delta W_p \tag{6.16}$$

$$\Delta W_t = \Delta\sigma\Delta\varepsilon_e + (\Delta\sigma\Delta\gamma_p)/(1+n) \tag{6.17}$$

Where n is the strain hardening exponent in the cyclic stress–strain curve described in Eq. (6.14). Similarly,

$$\Delta J = 2\pi a(\Delta W_e + \Delta W_p) \tag{6.18}$$

Components rarely experience stresses or strains that vary in a continuous and regular manner. There are generally periods in which stress or strain levels are fairly constant, such as power-on, operate, and power-off of an electronic device. During strain-controlled fatigue, the presence of a hold period is associated with stress relaxation—or fall-off. Dwells may exist at any position on the hysteresis loop but are usually applied at the strain extremes. Under certain conditions, the presence of a dwell can be profoundly influential on fatigue endurance [5]. This situation is often described as a fatigue–creep interaction, although a physical interaction between the damage processes is the exception rather than the rule [5]. Similar problems may arise when the strain rates in the cycle are unequal. For example, slow strain rates in tension followed by rapid deformation in compression produces shorter lives than when the strain–time profile is reversed.

6.2.4 Thermomechanical Fatigue

The origins of the forces in fatigue may be mechanical, such as external loads or fluctuating vibrations, or thermal. In the latter instance, the stresses and strains develop due to differing amounts of expansion or contraction during temperature changes. This damage may build up on a microscale, when different phases within the microstructure have different expansion coefficients, or on a macroscale when materials with different coefficients of thermal expansion are joined to each other, as in the case of the solder joint. The process is thermomechanical fatigue. This is the major mechanism of failure in service, particularly for surface mount systems. Strictly speaking, there are two forms of this process. The application of fluctuating temperatures to a body in which internal constraints to free expansion exist (such as different phases in a microstructure) is known as thermal stress fatigue (TSF). In contrast, a body exposed to temperature fluctuations, and for which free expansion is limited by external constraints, experiences thermomechanical fatigue. A key difference is that in TSF analysis of the stress–strain conditions is difficult, whereas in TMF these parameters may be independently controlled, making analysis relatively straightforward. Very often in the literature, this distinction is not made, and the term "thermomechanical fatigue" is used universally. The process is now examined in more detail.

During continuous cycling under strain control at a constant temperature, the complete stress–strain relationship may be represented by a hysteresis loop symmetrical about origin. The insertion of a dwell period in a cycle results in an asymmetric loop. Strain and temperature may be varied in-phase (peak temperature with peak strain) or out-of-phase (peak temperature with minimum strain). Idealized loops are shown in Fig. 6.7 and indicate how the TMF loop is essentially the product of the isothermal loops at minimum and maximum temperature. A compressive mean stress is generated for in-phase cycling, and a tensile mean stress for out-of-phase fatigue. Under conditions where there is significant cyclic plasticity, as is likely in solder alloys, mean stresses rapidly disappear and have little effect on endurance [6]. Figure 6.8 shows the hysteresis loops for a Sn–Ag eutectic solder at 25 and 80°C, and for TMF between those temperatures. In this example, the TMF profile lies between the two isothermal curves, but as will be seen later, the precise test conditions can have a substantial effect on its shape.

Figure 6.9 illustrates how the inclusion of a dwell period at maximum tensile strain in a cycle affects the hysteresis loop. During isothermal cycling, the stress increases progressively until the maximum strain limit is reached. On strain reversal, unloading occurs until the minimum strain is attained. For TMF, with in-phase cycling, the stress level may exhibit a maximum value and

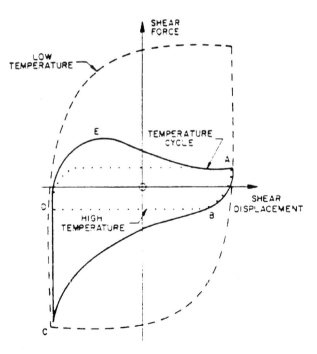

Figure 6.7   Evolution of a hysteresis loop produced by TMF from the isothermal loop at the temperature extremes. (From Ref. 34.)

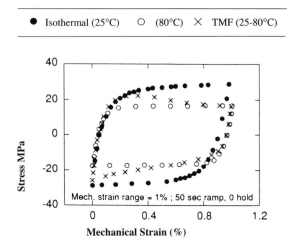

Figure 6.8 Comparison between isothermal and TMF hysteresis loops for an Sn–3.5Ag alloy. (From Ref. 35.)

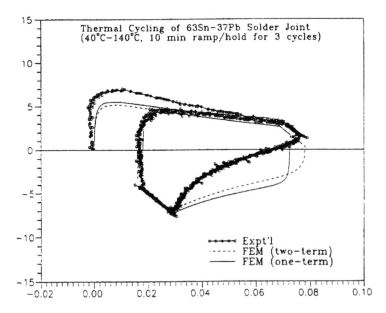

Figure 6.9 Characteristic TMF hysteresis loops for a cycle containing a dwell period. (From Ref. 36.)

then fall because the temperature dependence of strength means that at the higher temperature, a smaller stress is required to maintain the desired strain level. The extent of stress relaxation during the hold period is shown to be similar in each case. Because strength usually increases with diminishing temperature, a similar inflection is not apparent on strain reversal to low temperatures. The location and extent of inflections in the hysteresis loop are governed by the temperature sensitivity of strength, the work hardening characteristics and their temperature sensitivity, and the strain (or temperature) range of the test. With low strength, ductile materials such as solder, hysteresis relationships quite distinct from the classic loop may be observed.

6.3   MECHANICALLY RELATED FAILURE MODES

Printed wiring boards containing electronics components may fail by a variety of mechanisms. For design and life prediction to have a sound foundation requires a knowledge and understanding of the dominant process in any specific application. This section briefly describes failure modes that are governed by mechanical, rather than chemical or electrical, factors. Such information indicates the properties that are influential in performance and that must be fully evaluated.

Overload failures occur when the applied stresses exceed the yield or ultimate strength of ductile materials or the fracture strength of brittle materials. For the former, the effect of temperature is important because yield strength may fall substantially at elevated temperatures. Fracture of a brittle solid is dependent not only on the externally applied stresses but also on the defects it contains and their orientation. The salient property in this case is the fracture toughness, K_{IC}, of the material containing the defect. Brittle overload failure is promoted by high rates of loading as incurred by impact damage.

Prolonged application of stress or strain provides opportunities for time-dependent processes to occur. At the extremes, this results in either time-dependent extension, or when there is geometrical constraint, as, for example, in solder joints encapsulated within a ball grid array package, stress relaxation. Both events damage the material by grain boundary sliding, voiding, and cracking. A typical thermal profile of a PWB may involve ramps of varying temperature interspersed with periods in which the temperature remains constant.

There are two sources of failure of electronics equipment which experience fluctuating stresses or strains: (1) the low amplitude, high-frequency stresses produced by vibrations of parts of the equipment structure, and (2) the large amplitude, low-frequency strains caused by either local (power switching) or external temperature changes. In the latter, TMF, significant differences in coefficients of thermal expansion between the materials comprising the interconnection may generate substantial shear strains. These produce bands of intense deformation which eventually become cracks that extend with continued cycling until failure occurs. The high-frequency situation is usually elastically dominated and results in failure by the initiation and growth of a single crack.

6.3.1 Development of Strains During Thermal Cycling

It has been mentioned previously that a factor of five or more is possible between the coefficients of thermal expansion, α, of the materials involved in solder joints. When two materials are in contact, either on a microscale, such as a second phase particle in a matrix, or on a macroscale, as in a soldered joint, a change in temperature will produce different displacements in each material. A strain will be generated. Considering two pieces of material as shown in Fig. 6.10 and assuming that equilibrium (zero stress, zero strain) exists at temperature, T_o, a temperature change, ΔT, will induce a strain, ε_T, given by:

$$\varepsilon_T = \Delta T(\alpha_1 - \alpha_2) \tag{6.19}$$

Where α_1 and α_2 are the coefficients of thermal expansion of the two materials. Even with large differences in coefficients of thermal expansion (say 20×10^{-6}

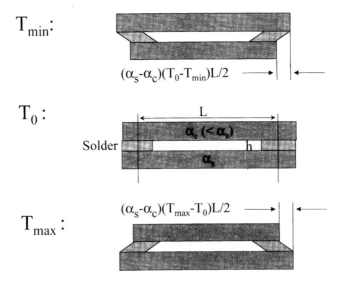

Figure 6.10 Development of strains during thermal cycling (schematic).

K^{-1}), and a value of ΔT of 100°C, the resulting strain would be fairly small ($\approx 0.2\%$). However, in an actual solder joint there is a "leverage" factor of the material between the solder joints. Fig. 6.10 illustrates this with a model joint comprising two bars of a material joined together by solder, thickness h, and separated by a distance, L. Here, α_s and α_c are the coefficients of thermal expansion of the solder and the "component", respectively; T_o, T_{min}, and T_{max} are the equilibrium, minimum, and maximum temperatures. If, as is likely, $\alpha_s > \alpha_c$, the solder contracts more on cooling and expands more on heating. The respective relative displacements are:

$$(\alpha_s - \alpha_c)(T_o - T_{min})L/2 \quad \text{on cooling} \tag{6.20}$$

and

$$(\alpha_s - \alpha_c)(T_{max} - T_0)L/2 \quad \text{on heating} \tag{6.21}$$

As the shear strain, γ_t, is displacement/height, then for a temperature excursion to T_{min}:

$$\gamma_t = \frac{(\alpha_s - \alpha_c)(T_o - T_{min})L}{2h} \tag{6.22}$$

Similarly, for a complete thermal cycle between T_{min} and T_{max}, the shear strain range, $\Delta\gamma_t$, is given by:

$$\Delta\gamma_t = \frac{(\alpha_s - \alpha_c)(T_{max} - T_{min})L}{2h} \tag{6.23}$$

This exceeds the strains developed in the earlier example of two blocks of materials in total contact by a factor of $L/2h$. According to their value, this can result in strains of several percent. Practically, the joint height may be very influential. Because the elastic range of most metallic materials is rarely greater than 0.1–0.2%, this means that temperature cycling can easily produce high strain fatigue conditions, in which the plastic strain range, $\Delta\gamma_p$, is much greater than the elastic strain range component, $\Delta\gamma_e$.

6.4 MEASUREMENT OF MECHANICAL PROPERTIES

The previous sections have demonstrated why reliable data regarding appropriate mechanical properties are essential for modeling, life prediction, and design purposes. It is unfortunate that much of the information that is available in the literature is either incomplete or relates to circumstances unlikely to be found in service.

A soldered interconnection is a quite complex structure. It comprises different materials with significant variations in properties between them, and the microstructure of the solder alloy is not only unstable and highly sensitive to thermal history, it contains intermetallic compounds (such as β-Cu_6Sn_5 and γ-Cu_3Sn) which can have a profound effect on performance. If the IMC layer becomes too thick, it can have an embrittling effect on the joint, and the strength and reliability are reduced. Additionally, a soft material (the solder) can withstand higher stresses if it is constrained by a harder one, i.e., the solder joint [7].

With regard to soldered interconnections, there are broadly three size scales that have been explored.

- Bulk solder properties have been determined on samples that are massive in comparison to an actual soldered joint. Stressing has generally been in tension and the effect of intermetallic layers ignored. However, stress and strain measurement and distribution are readily achieved and the resulting constitutive equations describe well the behavior of the particular alloy in its specified condition. The properties may be regarded as more "constant" than those determined from measurements on joints themselves and therefore more useful in analysis and modeling procedures. Nevertheless, they remain susceptible to temperature, strain rate, and microstructural instability effects.
- Model joints, typified by parallel cantilever bar or pin in ring specimens, enable shear stress and strain to be measured indirectly or calculated. They also incorporate the effects of the intermetallic and their size is nearer to that of real joints. Naturally, their performance is more geometry-specific than that of the bulk material.
- Actual joints usually have irregular shapes and experience complex stressing, although it is predominately in shear. The dimensions of the intermetallic layer may be significant in comparison with the solder thickness, and, consequently, the mechanical behavior of a particular type of interconnection becomes highly joint specific.

A key question, which remains unresolved, is the correlation, if any, between the bulk solder properties and joint performance. From the descriptions above, intuition suggests that no relationship should exist, although there are some experimental data available that suggest otherwise. For example, a good correlation between the torsional fatigue lives of bulk and pin-in-ring model joint specimens has been demonstrated provided that the latter contain no defects [8].

6.4.1 Testing Methods

In electronic equipment, boards may experience bending or twisting, while solder joints are generally sheared. The correlation between properties measured in tension and shear involves Poisson's ratio, v (which is the ratio of extension to lateral contraction in a tensile test). For example, the shear strength, γ_u, is related to the tensile strength, σ_u, by

$$\gamma_u = \frac{\sigma_u}{2(1+v)} \tag{6.24}$$

and the shear modulus, G, is related to the tensile modulus, E, by

$$G = \frac{E}{2(1+v)} \tag{6.25}$$

The most elementary means of applying a force to a specimen is by dead weight loading either directly or via a lever mechanism. Load increases are generally by step increments, so this type of machine is restricted to creep testing when a fixed load is applied and remains constant. For the application of a gradually increasing force, screw-driven machines have been traditionally employed for monotonic and cyclic testing. With such equipment, displacement rates are limited, especially during fatigue testing when a change in the direction of displacement is associated with delays and "flexibility" of the loading system. There are also substantial restrictions on the shape of the displacement–time cycle. These shortcomings are largely eliminated in test machines incorporating closed-loop servohydraulic control systems, and these are now regarded as standard for toughness, high strain fatigue, and crack propagation studies. For high cycle fatigue or crack growth studies at low stress ranges, resonance machines have been extensively used. However, their operating range and specimen geometry are limited due to the requirement for resonance to occur. The incorporation of prolonged dwell periods (several hours) in strain–time cycles requires the use of electrohydraulic machines because of their greater stability.

A test machine may be operated in one of three common control modes. Selection of the one most appropriate to the conditions experienced in service is desirable. Position control involves adjusting the position of the crosshead or loading actuator between predetermined limits, and by calibration or assumption, calculating the strains within the gauge length of the specimen. A load cell within the loading train provides stress values. Load control utilizes the load cell to apply loads between desired limits, which can be converted to stress by consideration of the cross-sectional area of the specimen. In strain control, a specific gauge length is selected as the operating range of an extensometer attached to the specimen. Displacements of the extensometer provide the required strain levels to enable direct strain control to be achieved. During mechanical or thermomechanical fatigue of a solder joint, the strains developed are usually concentrated in the solder, which is often the softest material. In effect, the solder is constrained by the adjacent materials, and the situation may be described as being under strain control. Therefore this mode of control is preferred for fatigue testing of solder alloys, especially when dwell periods are included in the strain–time cycle.

6.4.2 Failure Definitions

While a malfunction in electronic equipment is generally associated with a loss of electrical continuity, from the evaluation standpoint a mechanical criterion for failure is usually preferred. Complete fracture or separation is absolute, but it causes problems in testing and has no element of conservatism built in. Measurements of electrical resistivity are not particularly sensitive to dimensional changes, in a situation where other factors, such as phase coarsening and localized plasticity, are occurring. A more popular approach is to examine the stress changes required to maintain the constant strain range limits of the test. When cyclic hardening occurs (as in lead-rich, Sn–Pb solders), the onset of load fall-off after the saturation plateau is a convenient identification for failure. The situation for alloys that exhibit cyclic softening (such as eutectic lead–tin, or many of the lead-free compositions) is more complex. According to the applied strain range, a saturation plateau may, or may not, be apparent in the stress range vs. cycle plot. It is usual to adopt a percentage load (or stress range) fall-off from the initial cycle. Values such as 10%, 20%, or 50% are common. Which is selected may have a noticeable effect on the Coffin–Manson, or equivalent, graph (Fig. 6.11), and the

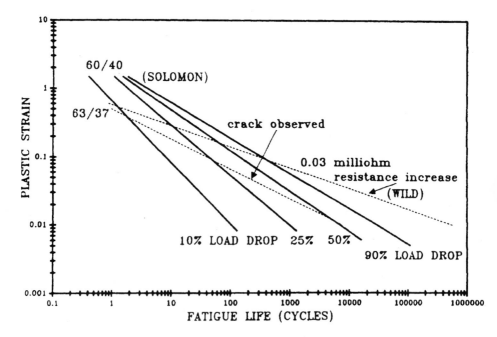

Figure 6.11 Effect of failure definition on fatigue life, depicted on a Coffin–Manson plot. (From Ref. 9.)

correlation between the failure definition and fatigue damage is unclear. However, during the fatigue of a Sn–40Pb alloy in shear, Solomon [9] has established the relationship between load drop and the extent of the cracked area. An alternative method is to identify the point at which crack growth begins to accelerate. This is achieved by plotting the stress range drop per cycle, $d\Delta\sigma/dN$, vs. number of cycles, when a clear transition in slopes results. A similar transition may be obtained by plotting the unloading modulus, $d\sigma/d\varepsilon$, immediately after a strain reversal as a function of cycles.

An alternative approach to defining a single failure criterion involves the mechanical assessment of damage. For example, a torque test or a shear displacement test on a joint provides a comparative measure of its remaining strength after cycling. Indirectly, this is indicative of the damage produced and the loss of structural integrity. Lead pull-out forces have also been determined for this purpose.

Continuous or interrupted surface observation of crack lengths provides no evidence regarding the extent of crack penetration. As both surface and penetrating cracks may be produced during TMF [10], this represents a serious limitation.

6.4.3 Standards for Mechanical Testing Solders

Many international standards exist which prescribe conditions under which mechanical properties should be determined (e.g., Ref. 11). This facilitates comparison of results from different laboratories. However, the high homologous temperatures encountered by solders in service are responsible for their low strength, high strain rate sensitivity, and significant microstructural instability. Data not obtained under identical test and material conditions are likely to exhibit a wider scatter than those for most other engineering alloys. The absence of uniformity in testing is largely responsible for the variability in mechanical property data reported in the literature. This problem is being addressed. For example, the Japanese Society of Material Strength (Solder Strength Working Group) has recently proposed detailed testing procedures for solders [12]. These recommend that bulk samples should be annealed for 1 hr at $T_h = 0.87$ prior to measurement of tensile strength and low cycle fatigue behavior. While this treatment, which is equivalent to 5–10 years exposure at room temperature, overcomes the problems of constancy of initial microstructure and its subsequent stability, the coincidence with the as-cast microstructure in an actual joint is questionable. This committee also proposes that a minimum strain rate of 2×10^{-2} sec^{-1} be used for determination of tensile properties in order to avoid creep deformation. For the same reason, an offset of 0.02%, rather than the usual 0.2%, is suggested for determination of a proof stress value. An alternative definition of fatigue failure is proposed as a 25% fall in tensile stress from that observed at half life which necessitates iterative testing. A strain rate of 10^{-3} sec^{-1} is recommended as standard for cyclic tests.

Equivalent, but not identical, recommendations from the United States [13] suggest aging at 0.67 T_m for 16 hr prior to tensile testing and strain rates in the range 10^{-3} to 10^{-4} sec^{-1}. Clearly, a substantial amount of work is still to be done both to produce the necessary data for lead-free solders and to adopt a uniform approach globally. However, reconciliation and the adoption of a uniform standard will represent significant progress and eliminate much confusion.

6.5 MECHANICAL EVALUATION OF MODEL JOINTS

A model joint should simulate the most commonly encountered stress situations, be made under easily controlled conditions which are as similar as possible to those in practical soldering and, finally, give reproducible test results [14]. Some examples of popular joint designs are described below and illustrated in Fig. 6.12.

In a single lap joint, the load is applied in a direction parallel to the joint interfaces. However, because the two component pieces cannot both lie in the axis of the load, the solder is subjected only initially to pure shear, and then nonlinearity of loading produces rotational forces which bend the component pieces and produce peeling stresses. These joints are relatively easy to manufacture. Bending is eliminated in double lap joints by the use of three component pieces. However, the resultant stressing condition is not pure shear as the configuration simply compensates for nonaxial loading. The whole assembly is massive in comparison to the single lap joint and it is more difficult to solder the joints within a practicable time.

The butt joint is the simplest type of joint, in which solder film is stressed in tension at right angles to the joint interfaces. Any nonaxiality of loading leads to peeling stresses and hence a reduced value for strength. The risk of this increases with joint area, and therefore small rods, with cross-sectional areas not greater than 50 mm^2, are preferred. The thickness of the solder layer varies between 0.1 and 3 mm. This type of joint is relatively difficult to manufacture, particularly for the larger joint gaps. Any final machining required could affect its properties.

The ring and pin joint is considered to be more representative of those found in service, because it more closely gives pure shear stressing conditions. However, the solder is not in line with the loading axis and any bending stresses due to nonaxial loading are internally balanced, but not eliminated. By using a small diameter rod and a relatively small overlap area, the mass of an assembled joint may be kept low, so that the rate of heating in solder-

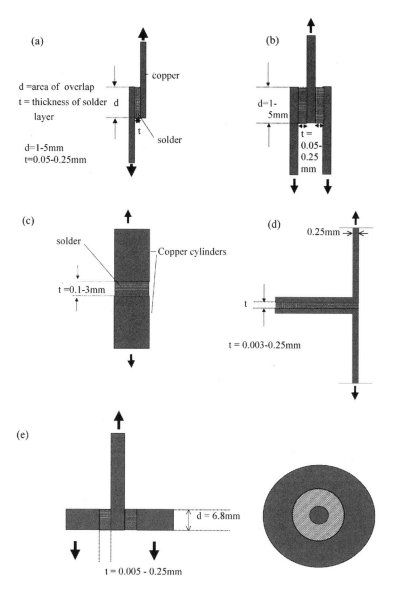

Figure 6.12 Various joint model joint geometries: (a) single lap joint, (b) double lap joint, (c) butt joint, (d) Chadwick (peel) test, and (e) pin in ring joint.

dipping is approximately the same as that in practical soldering applications. Strength values achieved by pushing the pin through the ring are slightly higher than those determined by pulling the pin due to the small radial expansion under compressive loading [14].

The Peel (Chadwick) Test is designed to measure the resistance to tearing or peeling of the interface in a solder joint. Narrow strips of metal are soldered parallel over a length, using aluminum foil spacers to control the joint gap. The two unsoldered ends of the component are bent away from each other to form a "T" shape and are pulled apart. This test effectively applies tensile stresses to the solder film at the point of peeling. Usually, an initial peak value, the load required for crack initiation, is ignored and the load required for crack propagation or peeling is recorded.

Other solder joint configurations have also been investigated due to their proximity to pure shear conditions. The Iosipescu joint is an accepted method for polymers and ceramics [15], and the asymmetrical four-point bend (AFPB) test has been favored because of its smaller specimen requirement [16].

Figure 6.13 illustrates the effect of different joint geometries, with a pin-in-ring joint [17] and a butt joint tested in an asymmetrical four-point bend test [18]. Shear strengths for the "butt" joint configuration are consistently

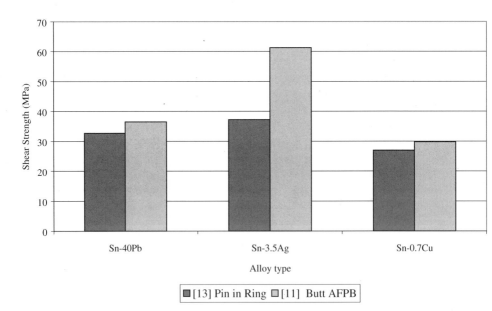

Figure 6.13   Variation of shear strength with joint geometry. (From Refs. 17 and 18.)

higher than those of the pin-in-ring joints, especially for the silver-containing alloy.

6.6   MECHANICAL EVALUATION OF ELECTRONICS ASSEMBLIES

The adaption of traditional mechanical test methods to electronics equipment requires careful attention with regard to specimen dimensions and to the magnitude of the forces involved. An actual solder joint could be 100 to 1000 times smaller than the bulk specimen that is being used to provide data on its mechanical properties (Fig. 6.14). Use of conventional test machines is

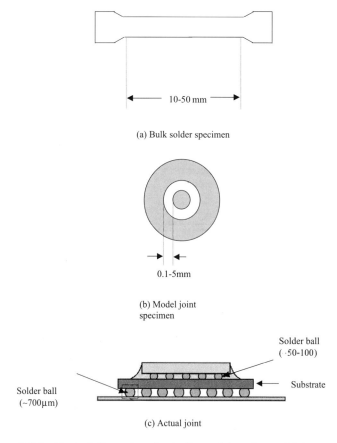

(a) Bulk solder specimen

(b) Model joint specimen

(c) Actual joint

Figure 6.14   Typical dimensions for bulk, model, and actual joint samples.

possible in all control modes although strain control becomes more difficult because of the dimensions of the specimen.

A point which manufacturers identify as a strong advantage of mechanical, rather than thermal, testing is the substantially shorter duration of the former. For example, an isothermal mechanical fatigue test at the same strain range as an equivalent thermal cycling test may be several hundred times shorter. However, this benefit merits close scrutiny. The problem of selecting which isothermal temperature to employ is compounded by the variation in influential mechanical properties with temperature.

With the recent increased interest in structural integrity in microelectronics, a new generation of test machines is evolving. The range of tests is considerable, and this is illustrated in Fig. 6.15. Shear tests can be performed on complete components or single solder balls. Layer peel tests and pull-out tests on individual wires can be carried out. Overall, much higher accuracy is necessary with respect to positioning, alignment, and force measurement. For example, the resolution and accuracy requirements on displacement are 50 nm and 2 μm, respectively. Temperature and environmental chambers may be added to the system, thus providing an extensive range of test conditions. A commercial system is shown in Fig. 6.16. A wide variety of joints, including gullwing, leadless (chip carriers), J-lead, plated through-hole, and

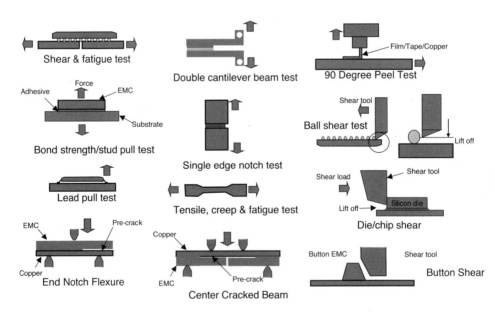

Figure 6.15 Overview of mechanical tests for microelectronic applications.

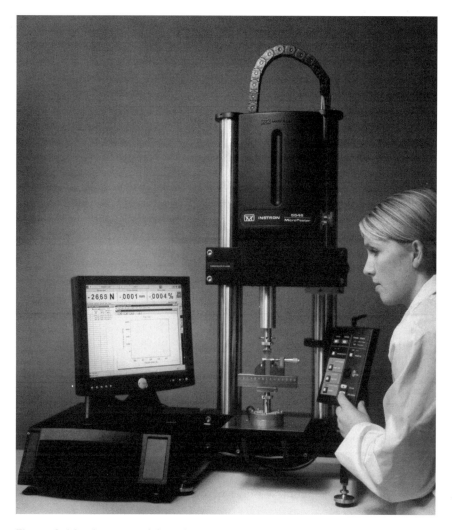

Figure 6.16 A commercial mechanical testing system for electronics applications coming next week.

ball grid arrays (BGA) is found in service. Each type has a different joint geometry, solder volume, and therefore properties. It is important to gain an understanding of how the geometrical and microstructural differences affect the strength and reliability of joints.

A new system capable of providing a rapid assessment of a wide range of mechanical properties, including TMF, of small specimens is the electro-thermomechanical test (ETMT) system [19]. DC heating produces a small central portion of the testpiece at constant temperature, and strain is determined from electrical resistance measurements. This method enables efficient discrimination between materials under similar conditions, but the inability to test under accurate strain control impairs its contribution to the development of reliable constitutive expressions.

6.7 THERMOMECHANICAL FATIGUE TESTING

Laboratory testing of actual components and structures for the purpose of life prediction presents a formidable challenge. There are very few instances where the component may be tested under real-time conditions employing actual stress–strain–temperature histories. In most cases, acceleration must be introduced into the test in order to furnish necessary information within an appropriate timescale. Furthermore, geometrical considerations, such as shape or scale, result in complex stress and strain distributions quite specific to the component being tested. However, stress analysis and modeling require generic information, which, in turn, necessitates evaluation of properties from quite simple specimens.

The conflict between accelerated testing of often complex, but realistic, specimens and idealized geometries from which values of stress and strain can more confidently be derived extends to the evaluation of solder joint performance. The results of such tests, say with regard to the effects of various parameters, should always be assessed in terms of the match between test conditions and the real service situation. In the following sections, the principal modes of thermomechanical fatigue testing will be presented together with some typical results.

6.7.1 Thermomechanical Fatigue of Bulk Specimens

Thermomechanical fatigue may be induced either by temperature cycling alone and allowing the cyclic stresses and strains to develop as a consequence of the geometry and the material, or by full external control of the temperature and strain cycle, in which bulk samples are subjected to externally applied cyclic strains and temperatures. This is the same as isothermal strain-controlled fatigue—with an additional temperature variation in-phase (IP) or out-of-phase (OP) with the applied strain. Heating is generally by resistance coils around the specimen or by RF induction, and cooling by forced air or liquid coolants. In addition to uniform straining, mechanical properties may be determined during testing by monitoring the stress response. When both temperature and strain are cycled, changes in specimen dimensions, thermally induced strain, and variation in mechanical properties as a function of temperature must be considered. These parameters change with temperature, so in order to test to strain limits, computer-controlled, closed-loop systems are essential.

As with all bulk testing, the correlation with actual joint behavior is uncertain, although this point has received little direct experimental attention. Similarly, as this form of test is complex and expensive, the extent to which the more basic isothermal strain-controlled fatigue is an indicator of TMF behavior merits more investigation. Current evidence is inconclusive.

The amount of data describing the fully controlled TMF of solders is very limited, but the shape of the hysteresis loop provides an insight into the key parameters in failure. In general, the stress range in TMF is greater than that for equivalent isothermal cycling [20] and the peak stresses are asymmetric about zero. Out-of-phase cycles are associated with a positive mean stress, and in-phase cycles result in a negative mean stress. Examination of the behavioral trends in other engineering alloys during TMF at similar homologous temperatures is indicative of possible responses in solders.

Steels and nickel alloys exhibit four characteristic patterns of behavior between fatigue life and the phase difference between temperature and strain [21].

1. $(N_f)_{IP} < (N_f)_{OP}$ at low strain ranges (low alloy steels—cast).
2. $(N_f)_{OP} < (N_f)_{IP}$ at low strain ranges (low alloy steels—forged).
3. $(N_f)_{OP} \approx (N_f)_{IP}$ for all strain ranges (stainless steel).
4. $(N_f)_{IP} < (N_f)_{OP}$ at high strain ranges (nickel superalloy).

The explanation for this wide, and sometimes apparently contradictory, behavior is simply that the dominant mechanisms of failure may change. For example, fatigue or creep may dominate singly, or they may interact. Fracture may be transgranular or intergranular. Individual mechanisms have different sensitivities to the test parameters and the material microstructure.

The effect of hold periods at maximum temperature has been shown to have little effect on endurance during out-of-phase cycling for steels [21,22], but a substantial reduction results for in-phase fatigue. These findings emphasize the need for an understanding of the prevailing dominant failure mechanisms.

The available results for solder alloys are generally coincident with those above. The presence of an in-phase thermal cycle substantially reduces the endurance of a Pb–3.5Sn solder with respect to that observed isothermally at the maximum temperature of cycling [23]. Contrary to most findings for isothermal fatigue, a reduction in frequency results in an increase in TMF life. Similar observations with regard to the comparison between TMF and isothermal fatigue life have been reported for Sn–37Pb, although the tensile portion of the TMF hysteresis loop and that for the maximum temperature were similar [24].

6.7.2 Thermal Cycling Model Specimens

This method simplifies the strain state by soldering together regular shaped pieces of materials with different coefficients of thermal expansion and exposing the joint to thermal cycling. Common combinations include stainless steel–solder–aluminum or Invar–solder–stainless steel. Polymers can also be included if they are plated. Due to the regular geometry and greater uniformity of temperature, stresses and strains can be calculated or measured directly, and a greater insight into microstructural instability may be achieved.

Pao et al. [25] have used model joints, comprising an alumina and a 2024-T4 aluminum alloy beam, to evaluate the thermal cycling performances of a range of solder alloys between 40 and 140°C. They assert that due to bending, such a configuration is neither under stress control nor strain control, and that both stress relaxation and creep occur during dwell periods. The life of the joint with the Sn–37Pb alloy was between 4 and 10 times greater than that with high lead alloy solder or lead-free solders, such as Sn–3Cu, Sn–4.0Cu–0.5Ag. None of the currently more popular lead-free compositions was examined. Using a three-beam (stainless steel-Invar-stainless steel) specimen for improved symmetry, Plumbridge and Liu [26] have established the cyclic stress–strain and Coffin–Manson relationships for Sn–37Pb. Associated metallography provides an insight into the responsible mechanisms.

6.7.3 Thermal Cycling Components/Boards

The significant benefit of this approach is that it utilizes actual specimens, and that deterioration may be monitored continuously, for example, by electrical means. Numerous specimens can be tested simultaneously which facilitates statistical analysis. However, it should be appreciated that this type of evaluation is empirical and provides highly specific information to the component (PWB)–temperature profile combination being tested. Table 6.1 illustrates some of the possible variables involved and demonstrates the enormous difficulty of the sole reliance on this approach, even for apparently minor alterations in operating conditions.

A range of temperature–time profiles, usually containing dwell periods, is applied, often at a high level (100 to 1000 times) of acceleration over the service situation. The thermal system may involve thermal shock by immersion in hot and cold baths, temperature cycling in an environmental chamber or power cycling (simply turning the device on and off). Clearly, there will be

Table 6.1 Summary of Possible Variations in Thermal Cycling PCBs

Variables in empirical testing	
Board	– dimensions, materials
Component	– type, size, number, arrangement
Service parameters	– T_{max}, T_{min}, t_{max}, t_{min}, rates of heating/cooling
	– location and duration of dwell periods
Interconnection	– process history, microstructure
What if service conditions change?	

significant differences in the strain rates, temperatures, and temperature gradients between these methods, and such differences may affect material behavior. For example, it has been shown that power cycling is more damaging than chamber cycling for chip carrier assemblies [27]. However, it is important to note the highly specific nature of such findings. Failure generally occurs as a consequence of the balance between competing mechanisms, and a small change in either temperature range or maximum temperature can produce a different failure mechanism.

Acceleration is generally achieved by a substantial reduction in the dwell duration in the temperature–time profile, and raising heating and cooling rates. Extrapolation to real-time conditions is hazardous, and many of the reasons for uncertainty have not yet been appreciated by the electronics sector. It is in this area, especially, where a thorough appreciation of materials behavior is required, and where much can be learned from other, more established, high-temperature design areas. For example, a faster strain rate may produce higher strength and increase the propensity for sudden catastrophic fracture. At low strain rates, deformation might be accommodated within the material's grains by anelasticity and plasticity. But at very low strain rates, grain boundary processes are likely to dominate. Each mechanism above has its own specific constitutive equation. Similarly, when acceleration is achieved by curtailing dwell periods, stress relaxation is restricted and the damaging intergranular processes at very low strain rates may be eliminated or curtailed. This simple example indicates that strain rate can have different, and even opposite, effects according to its value and the controlling deformation mechanism (Fig. 6.17). Until reliable deformation or fracture mechanism maps are available for materials in interconnections, the empirical approach will be required.

Joints may be regarded in two categories: leaded and leadless, i.e., with and without leads. Although the leadless chip carrier offers many benefits, it does depend totally upon a column of solder to perform satisfactorily over a wide range of conditions. The introduction of a compliant lead between the

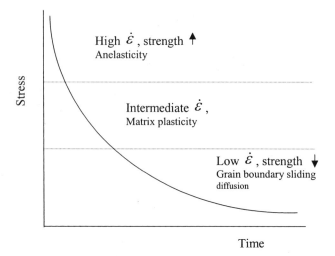

Time

Figure 6.17   Schematic illustration of the effect of strain rate on damage mechanisms and strength.

chip carrier and the substrate significantly reduces the thermal mismatch problem between the package and board, and the plastic strain range experienced by the solder. However, the elasticity of the leads may prolong the duration of creep stresses on the solder and induce elements of creep strain which are additional to the cyclic mismatch plasticity. During thermomechanical fatigue, creep ratcheting can occur due to the strong temperature dependence of creep during different phases of the loading cycle. More rapid and extensive creep occurs on the higher temperature side, whereas on the low temperature side the solder is more creep resistant and lead-deflection accommodates most of the strain. On reheating, the force from the deflected lead produces considerably more strain due to the higher temperature. Eventually, equilibrium is reached which is determined by the stiffness of the lead and the solder properties. While creep ratcheting is usually self-arresting, it may produce damage that accentuates subsequent fatigue processes. For example, it may cause rapid initiation of cracks.

The extent of empirical testing is governed by several factors: the desired service life, the level of acceptable failure—normally expressed as ppm, and the severity of the application (commonly, nine categories are cited ranging from Consumer Electronics to Automotive Underhood, as the most severe). In some cases, particularly involving leaded components under severe conditions that require a low failure probability, the number of tests demanded for statistical rigor is prohibitively large [28].

In addition to thermally cycling full-scale boards, multiple but discrete soldered joints between panels may be cycled in shear under displacement control with external air heating and cooling [29]. Continuity monitoring enables failure in a particular joint to be identified. While this method meets the requirements of scale, uniformity of strain, stress measurement, and a realistic failure criterion, it is quite complex and time consuming and requires highly sensitive strain resolution. It cannot be emphasized too strongly that all data are limited to the conditions under which they were derived. Both in terms of relevance to actual performance under service and to comparability (or ranking) of alloys, extrapolation of the findings to a broader spectrum is unsubstantiated.

In an extensive comparability program [30,31], joints comprising Sn–3.8Ag, Sn–0.7Cu, and Sn–3.8Ag–0.7Cu (and minor modifications of them) were subjected to a series of thermal cycling tests with Sn–40Pb and Sn–62Pb as comparator alloys. No significant difference in terms of performance under fluctuating temperatures was found between lead-containing and lead-free alloys. Figs. 6.18 and 6.19 demonstrate the broad similarities in performance for the thermal shock and power cycling. The ternary Sn–3.8Ag–0.7Cu was identified as the most likely universal replacement alloy.

Kariya and co-workers [32,33] have demonstrated the effect of copper and bismuth additions to an Sn–3.5Ag solder joint on a QFP assembly. In terms of the lead pullout strength, 0.5% and 1.0% Cu resulted in an improvement over those in the binary alloy (Fig. 6.20), which gradually reduced with thermal cycling between 0 and 100°C. While bismuth additions up to 7.5% were initially beneficial, subsequent thermal cycling caused a pro-

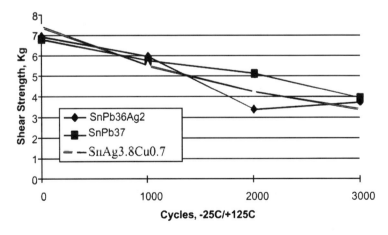

Figure 6.18 Effect of thermal cycling on joint shear strength. (From Ref. 30.)

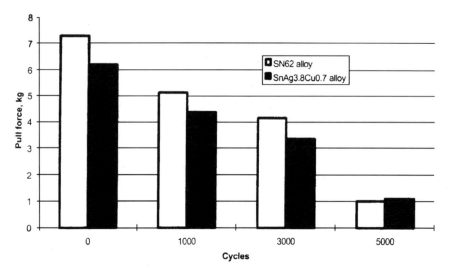

Figure 6.19 Effect of power cycling on joint strength. (From Ref. 30.)

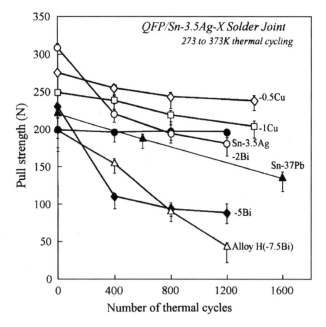

Figure 6.20 Effect of Bi and Cu additions on the TMF performance of Sn–3.5Ag alloys. (From Ref. 32.)

nounced fall in pullout strength to less than a half of that in the binary alloy and its value prior to thermal cycling. Pullout strengths in the binary Sn–3.5Ag were unaffected by up to 1200 cycles, while those in the lead–tin eutectic were reduced by some 30% by the same number of cycles.

This diversity of methods for assessing thermomechanical fatigue exacerbates the problem of comparability of results between laboratories. While the popular thermal cycling experiments involving complete boards provide valuable empirical information regarding the effect of specific parameters, their generic contribution is limited. Fully controlled thermomechanical fatigue is a complex process especially with regard to frequency effects. To model it reliably requires materials property data over the whole temperature range and under similar strain rates.

6.8 CONCLUDING COMMENTS

The fragmented nature of the electronics industry sector, as compared with power generation or aerospace, is responsible for the plethora of technologies in current use. Even small differences in practice can render comparisons between various laboratories and communication between companies a considerable challenge. Although mechanical properties and structural integrity are relative newcomers to the electronics arena, their importance is growing, with continued miniaturization and the appearance of the new generation of lead-free solder alloys. The shortcomings above apply equally to this area also.

Empirical testing of assemblies provides valuable information but adds little to the generic understanding of potential failure processes. Eventually, this approach, which constitutes the bedrock of present-day practice, will become incapable of providing sufficient reliability data within the time scales necessary to maintain competency. The goal of a paper-based design capability will require soundly based models that reflect service conditions and the availability of appropriate mechanical properties obtained under realistic conditions. These properties and the methods of their evaluation described in this chapter will assume increasing significance in the future.

ACKNOWLEDGMENTS

The author wishes to gratefully acknowledge the assistance received from his colleagues in the Solder Research Group at the Open University. In particular, the contributions of Shellene Cooper (née Peters), Yoshi Kariya, Xian Wei Liu, and Colin Gagg were particularly valuable.

REFERENCES

1. GR Blackwell. Circuit boards. In: GR Blackwell, ed. The Electronic Packaging Handbook. Boca Raton, FL: CRC Press, 2000, pp 5-1–5-29.
2. WJ Plumbridge, CR Gagg, S Peters. Creep of lead-free solders at 75°C. J Electron Mater 30:1178–1183, 2001.
3. Z Guo, H Conrad. Fatigue crack growth rate in 63Sn37Pb solder joints. J Electron Packag 115:159–164, 1993.
4. C Kanchanomai, Y Miyashita, Y Mutoh. Low cycle fatigue behaviour and mechanisms of a eutectic Sn–Pb solder 63Sn/37Pb. Int J Fatigue 24:671–683, 2002.
5. WJ Plumbridge. Metallography of High Temperature Fatigue: Properties and Prediction. In: RP Skelton, ed. London and New York: Elsevier Applied Science, 1987, pp 177–228.
6. BI Sandor. Fundamentals of cyclic stress and strain. Wisconsin: University of Wisconsin Press, 1972, pp 176–180.
7. JR Griffiths, JA Charles. The strength of soldered joints. Met Sci J 2:89–92, 1968.
8. M Kitano, T Shimuzu, K Tetsuo, Y Ito. Statistical fatigue life estimation: the influences of temperature and composition on low-cycle fatigue of Sn–lead solders. Curr Jpn Mater Res 2:235–250, 1987.
9. HD Solomon. Low cycle fatigue of 60/40 solder-plastic strain limited vs displacement limited testing. Electronic Packaging: Materials and Processes. Am Soc Met, 1986, pp 29–49.
10. XW Liu, WJ Plumbridge. Damage produced in model eutectic Sn–Pb joints during thermal cycling. J Electron Mater 32:278–286, 2003.
11. ASTM Standards. ASTM E606: standard practice for strain-controlled fatigue testing, vol. 03.01. American Society for Testing and Materials, 1998, pp 525–539.
12. M Sakane, H Nose, M Kitano, H Takahashi, M Mukai, Y Tsukada. Tensile and low cycle fatigue standard testing for solders—JSMS Recommendation. Tenth International Conference on Fracture, Hawaii, Dec 2001.
13. C Handwerker. Workshop on Modelling and Data Needs for Lead-Free Solders', TMS Annual Meeting, New Orleans, February 2001.
14. CJ Thwaites, R Duckett. Effects of soldered joint geometry on their mechanical strength. Rev Soudre 4:1–6, 1976.
15. ASTM Standards D5379-93 and C1292-95a. Philadelphia: American Society For Testing Materials, 1995 and 1997.
16. Ö Ünal, DJ Barnard, IE Anderson. A shear test method to measure shear strength of metallic materials and solder joints using small specimens. Scr Mater 40:271–276, 1999.
17. JC Foley, A Gickler, FH Leprovost, D Brown. Analysis of ring and plug shear strengths for comparison of lead-free solders. J Electron Mater 29:1258–1263, 2000.
18. IE Anderson, TE Bloomer, RL Terpstra, JC Foley, BA Cook, J Harringa.

Development of eutectic and near eutectic Sn–silver–copper solder alloys for lead-free assemblies. IPC Works 1999: An International Summit on Lead-Free Electronics Assemblies, Minneapolis, 1999.

19. B Roebuck, MG Gee, M Brooks, D Cox, R Reed. Miniature multiproperty tests at elevated temperatures. In: WJ Evans, RW Evans, MR Bache, eds. COMPASS99 Component Optimisation from Materials Properties and Simulation Software. Swansea: University of Wales, 1999, pp 233–240.

20. K Kuwabara, A Nitta, T Kitamura. Thermal–mechanical fatigue life prediction in high temperature component materials for power plant. In: DA Woodford, JR Whitehead, eds. Proc. International Conference on Advances in Life Prediction Methods. New York: ASME, 1983, pp 131–141.

21. K Kuwabara, A Nitta. Isothermal and thermal fatigue strength of Cr–Mo–V steel for turbine rotors. Central Research Inst. of Electric Power Industry (CRIEPI) Report, E277005, 1977.

22. A Nitta, K Kuwabara. Thermal–mechanical fatigue failure and life prediction. In: R Ohtani, M Ohnami, T Inoue, eds. High Temperature Creep-Fatigue. London and New York: Elsevier Applied Science, 1988, pp 203–222.

23. LR Lawson, ME Fine, DA Jeanotte. Thermomechanical fatigue of a lead alloy. Metall Trans 22A:1059–1070, 1991.

24. EC Cutiongco, DA Jeannotte, ME Fine. Isothermal fatigue behaviour of 63Sn–37Pb solder. J Electron Packag 112:110–114, 1990.

25. YH Pao, S Badgley, R Govila, E Jih. Thermomechanical and fatigue of high-temperature lead and lead-free solder joints. In: SA Schroeder, MR Mitchell, eds. Fatigue of Electronic Materials. ASTM Spec Tech Publ vol. 1153. Philadelphia: American Society for Testing and Materials, 1994, pp 60–81.

26. WJ Plumbridge, XW Liu. Thermomechanical fatigue behaviour of Sn–37Pb model joints. Matls Sci Engrg. To be published.

27. J Lynch, A Boetti. Solder joint reliability of leadless chip carriers. In: JH Lau, ed. Thermal Stress and Strain in Microelectronics Packaging. New York: Van Nostrand Reinhold, 1993, pp 579–606.

28. W Engelmaier. Solder attachment reliability, accelerated testing, and result evaluation. In: JH Lau, ed. Solder Joint Reliability—Theory and Applications. New York: Van Nostrand Reinhold, 1991, pp 545–587.

29. DR Frear. Thermomechanical fatigue of solder joints: a new comprehensive test method. IEEE Trans Components Hybrids Manuf Technol 12:492–501, 1989.

30. M Warwick. Implementing lead-free soldering—European consortium research. Proc. Surface Mount Technology Association Journal of Surface Mount Technology—Fifth Annual Pan Pacific Microelectronics Symposium, 2000, pp 1–12.

31. MR Harrison, JH Vincent, Y Steen. Lead-free soldering for electronics assembly. Solder Surf Mt Technol 13:21–38, 2001.

32. Y Kariya, Y Hirata, M Otksuka. Effects of thermal cycles on the mechanical strength of quad flat packleads/Sn–3.5Ag–X (X = Bi and Cu) solder joints. J Electron Mater 28:1263–1269, 1999.

33. K Warashina, Y Kariya, Y Hirata, M Otsuka. Effect of hold time on the thermal fatigue damage of QFP/Sn–3.5mass%–X (X = Bi and Cu) solder joint. Sixth

Symp. on Microjoining and Assembly Technology in Electronics (MATE 2000), Yokohama, 2000, pp 275–280.

34. PM Hall. Forces, moments and displacements during thermal chamber cycling of leadless ceramic chip carriers soldered to printed boards. IEEE Trans Components Hybrids Manuf Technol 7:314–327, 1984.

35. H Mavoori, S Vaynman, J Chin, B Moran, L Keer, M Fine. Mechanical behaviour of eutectic Sn–Ag and Sn–Zn solders. Materials Research Society Symposium. vol. 390. Materials Research Society, 1995, pp 161–175.

36. YH Pao, S Badgley, R Govila, E Jih. Thermomechanical and fatigue behaviour of high temperature lead and lead-free solders joints. In: SA Shroeder, MR Mitchell, eds. Fatigue of Electronic Materials. ASTM Spec 1153. American Society for Testing and Materials: Philadelphia, 1994, pp 60–81.

7
Electrochemical Migration

Hirokazu Tanaka

ESPEC Corp., Tochigi, Japan

7.1   INTRODUCTION

This chapter will first look into the phenomenon of electrochemical migration, which affects the insulation characteristics of electronic components, and will also present a brief outline of the form and structure of migration buildup. Following that, the electrochemical characteristics of lead-free solder alloys will be considered, and the speed of anode dissolution and the form of cathode deposits during the process of migration buildup will be compared with the characteristics of conventional eutectic solder. Finally, this section will consider reliability tests for printed wiring board (PWB)-mounted components, always a major concern in commercial applications, and will discuss the electrical reliability of these lead-free solder alloys.

7.2   ELECTROCHEMICAL MIGRATION

7.2.1   Terminology and Research Trends

Electrochemical migration is a generic term applied in general to the migration of metal atoms and metal ions. In this chapter, the term "electrochemical migration" is used to indicate the electrochemical phenomenon in which substances such as moisture and dew condensation occurring between PWB electrodes cause metal ions to migrate from one electrode to the opposite electrode, producing a buildup of deposits composed of metal or chemical compounds.

This electrochemical migration is also referred to by such names as "metal migration" (1–3) and "ionic migration" (4,5), but "electrochemical migration" is the generally accepted term among industry groups such as the association connecting electronics industries (IPC) (6,7).

This chapter uses the term "electrochemical migration" to indicate the type of migration described above. Table 7.1 shows the types of migration in addition to electrochemical migration that occur in electronic components. Electro migration, stress migration, and thermal migration are all specific phenomena occurring inside the semiconductors, and this table presents only a brief summary of the meaning of each.

Experimental research into electrochemical migration (hereafter, migration) was first carried out in the United States in Bell Laboratories by Kohman et al. (8) and described in a paper published in the *Bell System Technical Journal*. This research involved the migration of silver from silver-plated terminals used on a telephone switchboard. The phenomenon was thought to have been caused by d.c. voltage and high humidity, and the report described experiments involving surface observation and changes in insulation resistance. As a result, silver migration was confirmed to cause degradation of insulation and serves as a factor affecting the long-term performance of electronic equipment.

However, this phenomenon occurred only in a specific environment with conditions of high humidity over a long period of time. Since there was not a large cluster of occurrences, the phenomenon was not taken up in further research. The development later in the 1970s of miniaturized equipment using PWBs led to more frequent degradation of insulation between electrodes on the PWBs resulting in failures, and so the research was pursued further at that time.

7.2.2 Migration Process Mechanisms

"Migration" refers to a phenomenon in which metal ions are transferred from one metal electrode to the opposite metal electrode. This process results in metal or metal-oxide deposits. The transfer occurs between the electrodes of

Table 7.1 Classification of Migration

Electrochemical migration: metals ionize and migrate due to an electric field.
Electro migration: interaction between metal atoms and electrons causes dispersion.
Stress migration: mechanical stress causes metal atoms to migrate.
Thermal migration: thermal stress causes metal atoms to migrate.

---

devices such as PWB when an electric field is impressed in the presence of moisture such as dew condensation adhering between the electrodes.

Migration is classified as either "dendrite" (6,7,9) or "conductive anodic filament (CAF)" (6,7,10) depending on the shape of the deposits and the conditions leading to the occurrence. "Dendrite" refers to the dendritic-shaped metal or metal-oxide deposits on the surface of PWB insulation. "Conductive anodic filament" refers to metal or metal-oxide

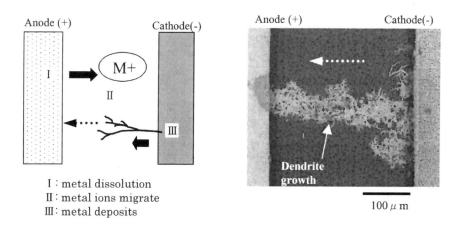

(a) Dendrite on the surface of PWB

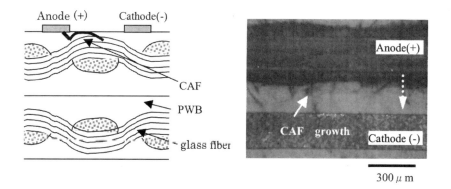

(b) CAF (Conductive Anodic Filament) of interior of PWB

Figure 7.1 Classification of electrochemical migration.

deposits in the shape of elongated fibers deposited along the glass fibers of the interior of the PWB insulation panel (Fig. 7.1).

Figure 7.2 shows the electrode reaction of the migration process for metal. There are three processes leading to migration: (I) anodic reaction (metal dissolution), (II) cathodic reaction (metal or metal-oxide deposits), and (III) interelectrode reaction (metal-oxide deposits):

(I) Anodic reaction ("M" stands for metal)

$$M \rightarrow M^{n+} + ne^- \tag{7.1}$$

$$H_2O \rightarrow 1/2O_2 + 2H^+ + 2e^- \tag{7.2}$$

$$M + H_2O \rightarrow MO + 2H^+ + 2e^- \tag{7.3}$$

(II) Cathodic reaction

$$M^{n+} + ne^- \rightarrow M \downarrow \tag{7.4}$$

$$O_2 + 2H_2O + 4e^- \rightarrow 4OH^- \tag{7.5}$$

$$2H_2O + 2e^- \rightarrow H_2 + 2OH^- \tag{7.6}$$

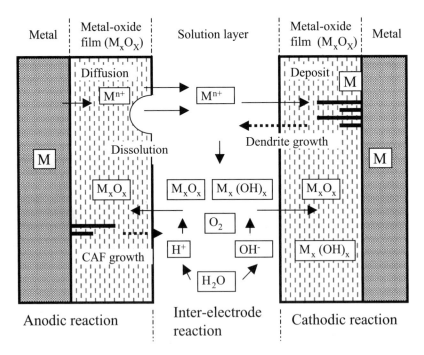

Figure 7.2 Electrode reaction of electrochemical migration process for metal. (From Ref. 5.)

(III) Interelectrode reaction

$$M^{n+} + 2OH^- \rightarrow M_nO \downarrow + H_2O \tag{7.7}$$

$$M^{n+} + 2OH^- \rightarrow M_n(OH)_2 \downarrow \tag{7.8}$$

Reactions (I) and (II) are thought to be mechanisms leading to dendrite formation. Metal dissolving at the anode (positive electrode) is deposited on the cathode (negative electrode). These dendrites consist of pure metal or metal oxide deposits growing toward the cathode and result in short circuits between the electrodes. Reactions (I) and (III) are all thought to be mechanisms leading to CAF formation. Here, too, metal dissolving at the anode results in metal or metal-oxide deposits between the electrodes, and the growth of these deposits results in short circuits between the electrodes.

Considering the mechanisms by the type of reaction, we find that the electrolytic reaction of water causes the area around the anode to become acidic due to H^+ ions forming from type (7.2) reactions, while the area around the cathode becomes alkaline due to OH^- ions forming from types (7.5) and (7.6) reactions (11).

Fig. 7.3 shows Pourbaix potential-pH diagrams (12) of the main materials used for lead-free solder. As you can see from the diagrams, when a pH shift toward acidity occurs, there is a greater tendency for metal ions to form, promoting a metal dissolution reaction at the anode. A shift toward an alkaline pH causes a metal deposition reaction at the cathode due to the formation of hydroxides and metal oxides. The reaction shown in formula (7.5) forms the driving force of the migration reaction. Oxygen gas, which makes up approximately 21% of the atmosphere, dissolves in the electrolytic solution, and because of the high oxygen reduction reaction (affinity for electrons) of this dissolved oxygen, electrons are captured at the cathode.

In this way, the process leading to migration is comprised mainly of anodic dissolution and cathodic deposition reactions, and the speed of those reactions is affected greatly by the characteristics of the materials.

7.2.3 Factors Affecting Migration

Migration results from the transfer of mass induced by such factors as charge transfer in solution, diffusion, and electrical migration. The reaction speed is affected by such factors as electrode potential of metal electrodes, solubility, stability of the passivity film, migration speed of metal ions, and fractal characteristics of deposits.

Table 7.2 shows the major factors affecting migration. The actual environment contains a wide variety of environmental factors (e.g., temperature, humidity, voltage, PWB moisture absorptivity, and halides in the flux) that affect these reactions and increase the reaction speed.

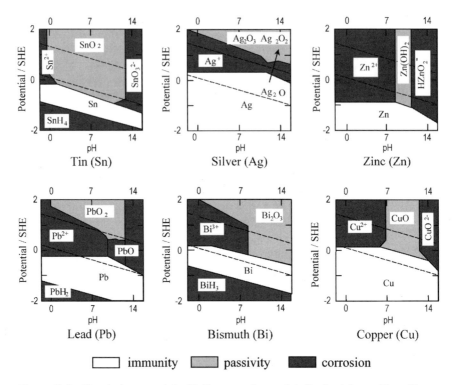

Figure 7.3 Pourbaix potential-pH diagram of materials for lead-free solder. (From Ref. 12.)

Table 7.2 Factors Affecting Migration

Factor	Effect on migration
Electrode material	(Greater tendency) Ag > Pb > Cu > Sn–Pb > Sn > Au (less tendency)
Temperature	The higher the temperature, the greater the acceleration of the reaction.
Humidity	The higher the humidity, the greater the amount of moisture absorbed, leading to acceleration.
Voltage	The higher the voltage, the greater the acceleration of the reaction.
Impurities	Halide impurities (e.g., Cl and Br) accelerate the reaction.
Type of substrate	The higher the moisture absorptivity of the material, the greater the acceleration (e.g., paper phenol PWB).

7.3 EVALUATION METHODS

Table 7.3 shows test standards and conditions for evaluating migration.

7.3.1 Simple Test Methods

Simple test methods (such as water drop test, such standards as IPC-TM650-2.6.13) make it possible to evaluate in a short time the tendency for migration to occur in different electrode metals. This is effective for comparing relative migration characteristics.

This method consists of dripping ion exchange water between electrodes and impressing d.c. voltage and then measuring the time to short-circuiting between the electrodes due to the occurrence of migration. Evaluation equipment is composed of a d.c. power supply, an ammeter, and a current limiter resistance. Observation of the migration is performed visually and with a microscope.

7.3.2 Environmental Test Method

Using environmental test equipment, d.c. voltage is impressed in the electrode gap, and the length of time to short-circuiting or insulation degradation is compared. Tests such as temperature–humidity–bias test (THBT) and the highly accelerated temperature and humidity stress test (HAST) are used to evaluate the moisture absorptivity of general materials. The temperature–humidity–cyclic test and the dew condensation cyclic test are used to evaluate factors caused by surface dew condensation. Measurement methods include methods such as the following:

1. Visual observation and microscopic observation of migration occurrence.
2. Measuring the insulation characteristics with an insulation resistance tester.
3. Checking the operation of the components being tested.

7.4 SURFACE STRUCTURE AND ELECTROCHEMICAL CHARACTERISTICS OF SOLDER ALLOYS

7.4.1 Surface Structure of Solder Alloys (13)

Alloys consisting of two or more elements dissolve and form chemical compounds with characteristics that differ from those of their individual components. Analysis of the migration process for alloy materials indicates that the following factors are involved in such reactions: (I) surface structure,

Table 7.3 Standards and Conditions for Migration Test

Standard number: date of issue	Title	Test conditions	Applied voltage	Measurement voltage
IPC-TM-650-2.6.13: 1985	Assessment of susceptibility to metallic dendritic growth: Uncoated printed wiring	room temperature; deionized water:	15 V/d.c. (max: 1.5 mA)	
ISO 9455-17: 2002	Surface insulation resistance comb test and electrochemical migration test of flux residues	40°C; 93% **RH**; 21 days 85°C; 85% **RH**; 21 days 40°C;	50 V/d.c. 50 V/d.c.	50 V/d.c. 50 V/d.c.
JIS-Z-3197-8.5.4: 1999	Test methods for soldering fluxes: migration test	90 to 95% **RH**; 1000 hr 85°C; 85 to 90% **RH**; 1000 hr 40°C; 91% ± 2% **RH**; 500 hr 65°C;	45 to 50 V/d.c. 45 to 50 V/d.c. 10 V/d.c.	100 V/d.c. 100 V/d.c. 100 V/d.c.
IPC-TM650-2.6.14.1: 2000	Electrochemical migration resistance test	88.5% ± 3.5% **RH**; 500 hr 85°C; 88.5% ± 3.5% **RH**; 500 hr	10 V/d.c. 10 V/d.c.	100 V/d.c. 100 V/d.c.

(II) electrode potential, (III) anode polarization, and (IV) stability of the passivity film that forms on the surface.

The main component of lead-free solder, Sn, is itself a metal with superb corrosion resistance due to the passivity film it forms. However, adding various alloy elements can degrade the corrosion resistance. In particular, verification can be considered necessary for such components as Ag and Cu, in which migration occurs relatively quickly.

The alloys Sn–Ag, Sn–Cu, and Sn–Bi currently being promoted for lead-free solder applications utilize the noble metals Ag, Cu, and Bi. Another alloy, Sn–Zn, utilizes the base metal Zn with Sn, and so that the conditions leading to migration in this alloy are likely to differ from those in the noble metal alloys. Figure 7.4 shows backscattered electrode image of the solder surface. Sn–3.5Ag shows the intermetallic compound (Ag–Sn) phase forming as a product of the reaction between Sn and Ag, with the Ag_3Sn phase dispersed throughout the Sn component. Similarly, Sn–0.75Cu shows the intermetallic compound (Cu–Sn) phase forming from Sn and Cu, with (Cu_6Sn_5) minutely dispersed throughout the Sn component. No intermetallic compound is formed in Sn–58Bi, Sn–9Zn, and in Sn–37Pb, and the photos show the second phase structure of Bi–Sn, Zn–Sn, and Pb–Sn.

The above evidence indicates that the surface structure of materials known as lead-free solder differs according to the composition of the alloy materials and their intermetallic compounds. These structures are affected by the electrode potential and the formation of a passivity film and appear as characteristics leading to the occurrence of migration.

7.4.2 Electrochemical Characteristics

Figure 7.5 shows the rest potential for the solder alloys and the single metals (Sn, Ag, Cu, Bi, Zn, and Pb) in a 0.1-mol aqueous solution of KNO_3. The horizontal axis shows the rest potential for the single metals, and the vertical axis shows the solder rest potential in relation to the rest potential for the additives in the solder. Figure 7.6 shows the current–potential curves for the solder alloys and the single metals (Sn, Zn, and Pb) in a 0.1-mol aqueous solution of KNO_3. In regard to anodic dissolution characteristics, these results show that an abrupt rise in current density during the sweep from low potential to high potential promotes the dissolution reaction.

A 0.1-mol aqueous solution of KNO_3 was used because the solution is neutral (pH = 6) and so does not promote the formation of solder compounds.

Sn–Ag and Sn–Cu: Sn–3.5Ag and Sn–0.75Cu, which form intermetallic compounds, exhibit approximately the same rest potential and solubility as Sn, and these solder alloys are affected by Sn. These properties indicate that Ag and Cu, which conventionally undergo a relatively fast occurrence

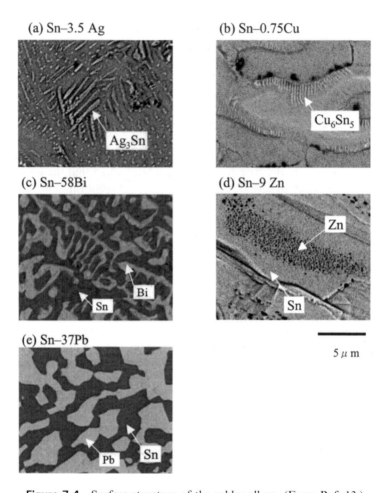

Figure 7.4 Surface structure of the solder alloys. (From Ref. 13.)

of migration, do not dissolve in solution due to intermetallic compounds formed with Sn (Ag_3Sn and Cu_6Sn_5) and so are affected by the characteristics of Sn.

Sn–Bi: Saturated Calomel Electrode: SCE. The rest potential of Sn–58Bi exhibits a more noble potential (-350 mV *vs.* SCE) than Sn. This is due to the hybridization potential of Bi and Sn that results from the high Bi content in the solder. The solubility of Sn–Bi exhibits approximately the same tendencies as Sn. The anode surface after migration testing (see Fig. 7.7) shows Bi remaining and Sn selectively dissolved out, and this indicates that Bi does not

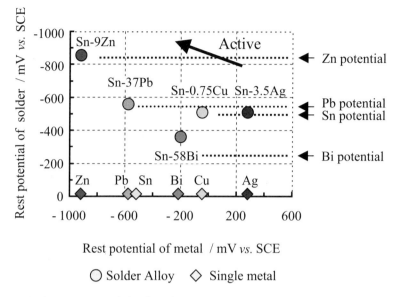

Figure 7.5   Rest potentials of various metals and solder alloys in 0.1 M KNO₃ aqueous solution. (From Ref. 5.)

dissolve in solution, and that Sn–3.5Ag and Sn–0.75Cu are both affected by the solubility of Sn.

Sn–Zn: Sn–9Zn exhibits a rest potential equivalent to the potential of Zn, indicating that this solder material is affected by the potential of Zn. The solder begins to dissolve in the vicinity of the solubility potential of Zn, approximately −700 mV vs. SCE, and then the solder exhibits an inflection point at approximately −180 mV *vs.* SCE, which is the solubility potential of Sn. After that, the solder exhibits the same solubility as Sn. This inflection point is thought to indicate the formation of a passivity film and is assumed to suppress the excess dissolution of Zn. From this, we see that the solubility characteristics of Sn–9Zn exhibit suppression of solubility of the base metal Zn due to the formation of a passivity film.

Sn–Pb: Comparing the rest potential of Sn–37Pb and Pb indicates that this solder is affected by the potential of Pb. This solder dissolves at the base potential (−350 mV *vs.* SCE) of Sn. According to the Pourbaix potential-pH equilibrium diagram, the corrosion area of the Pb component is greater than that of Sn, and so there is a greater tendency for the dissolution reaction to occur, and the Pb component dissolves out first. In this way, the electrode potential and the dissolution reaction related to the electrochemical charac-

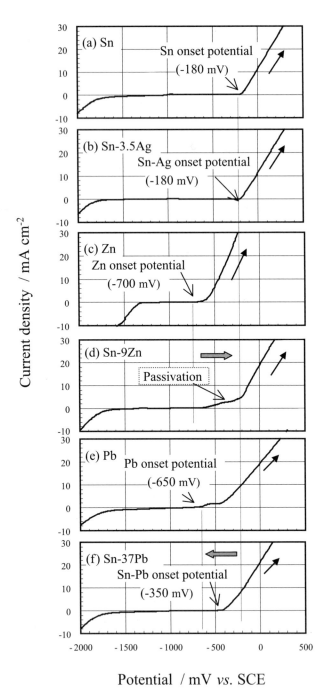

Figure 7.6 Current–potential curves (solubility) for various metals and solder alloys in 0.1 M KNO₃ aqueous solution. (From Ref. 5.)

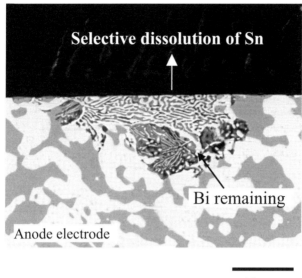

Figure 7.7   Anode electrode of Sn–58Bi solder after the migration test.

teristics of the solder compounds are affected by the solder composition and the surface structure of the solder.

Overall, Sn–Ag, Sn–Cu, and Sn–Bi exhibit the characteristics of Sn, while Sn–Zn and Sn–Pb exhibit the characteristics of Zn and Pb. The speed of the dissolution reaction is suppressed by the high stability of the passivity film and thus is slower than conventional Sn–Pb eutectic solder. Because of these results, subsequent reports will take up the differing electrochemical characteristics of Sn–Ag, Sn–Zn, and Sn–Pb and compare their migration processes.

7.5   MIGRATION PROCESS OF SOLDER ALLOYS

7.5.1   Analysis of the Anode Dissolution Process Using Electrochemical Techniques (5,14)

Figure 7.8 shows the relationship between resonance frequency variation and the quantity of electricity of the anode dissolution process according to the quartz crystal microbalance (QCM) method (4). Because Sn–3.5Ag shows solubility approximating that of the theoretical gradient of Sn^{2+}, Sn–3.5Ag is thought of as dissolved Sn^{2+}. In its initial reaction, Sn–9Zn shows the

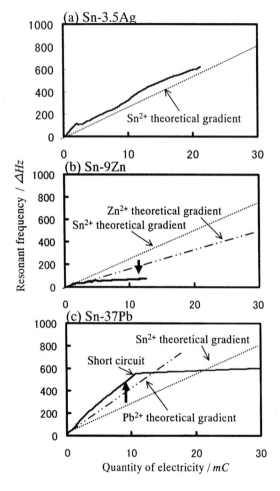

Figure 7.8 Relationship between quantity of electricity and resonant frequency (change in mass) of anode dissolution process by QCM method. (From Ref. 5.)

solubility of Zn^{2+}, but later, the speed at which it dissolved slowed markedly. This is thought to be the result of a passive condition stemming from something such as an oxide film at the surface of the electrode suppressing dissolution. Sn–37Pb conformed approximately to the theoretical gradient of Pb^{2+}, with Pb^{2+} taking priority in dissolving out. The speed at which each type of lead-free solder dissolved was slower than that of Sn–37Pb.

Figure 7.9 shows the equivalent circuit of the migration process inferred from the electrochemical impedance spectroscopy (EIS) method (15). Calcu-

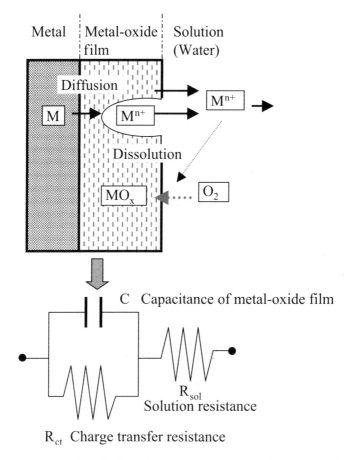

Figure 7.9 Proposed equivalent circuit of anodic dissolution process during the migration test. (From Ref. 14.)

lated from this equivalent circuit are the charge transfer resistance (R_{ct}), resonant frequencies, and time variability of the electrical current values, all shown in Figure 7.10. Looking at the time variability of R_{ct} and the resonant frequencies, we see an increase in anode weight concurrent with the drop of R_{ct}. In other words, the dissolved weight increases leading to short-circuiting between the electrodes.

Changes in R_{ct} indicate the ease of transit of metal ions and electrons at the electrode, and changes in the interface capacity are related to growth of an oxidation film (passivity film) caused by reactions with oxygen and to the formation of an electrical double layer in the solution. Results of the experiments indicate that Sn–3.5Ag and Sn–9Zn, which exhibited smaller changes

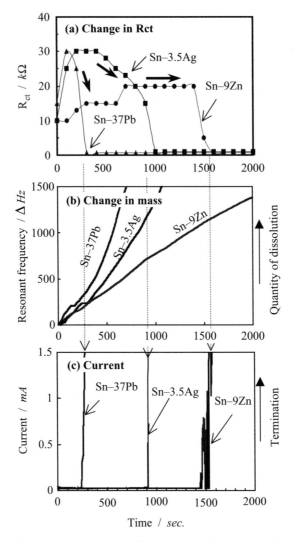

Figure 7.10 Change in R_{ct}, resonant frequency, and current during the migration process by EIS method. (From Ref. 14.)

in R_{ct} and interface capacity, have a slower reaction speed and the dissolution reaction is suppressed.

Comparing the various types of solder alloys indicates the following.

Sn–37Pb showed a sharp drop in R_{ct} immediately after the experiment started, with a concurrent increase in anode dissolution weight leading to short-circuiting, and so this material had sudden changes in R_{ct} and exhibited a fast migration time.

Sn–3.5Ag showed an extremely high spike in R_{ct} immediately after the experiment started, but this gradually tapered off. At that time, the anode dissolution weight leading to short-circuiting increased, and so because the R_{ct} change was comparatively mild, this material had a slow migration time.

Sn–9Zn experienced rather sluggish changes in R_{ct} and deposits after the experiment began. Later on, there was a sudden drop leading to short-circuiting. This induced the formation of a passivity film, which inhibited the charge transfer, and so this material is classified as an inhibitor of migration.

Figure 7.11 shows the results of anode cross-section observation following migration testing using the QCM method. With Sn–3.5Ag, an Ag_3Sn compound was left in a finely dispersed layer, and the Sn component was confirmed to be markedly dissolved and facing inward (16). With Sn–9Zn, there was evidence of shallow pitting corrosion, and from the EPMA analysis (Fig. 7.12), we see that a high Zn oxide-film layer formed at the surface. Sn–37Pb eutectic solder showed overall corrosion and, in particular, the Pb component dissolved out first.

The above results were used to examine the causes of migration for the various solder alloys. Sn–3.5Ag exhibited a high R_{ct} resulting from the formation of a passivity film layer of materials such as SnO_2 at the electrode surface. The comparatively gradual breakdown of that film layer produced relatively slow migration growth.

Sn–9Zn forms passivity layers from such materials as ZnO, $Zn(OH)_2$, and SnO_2, and due to the long time that this passivity layer maintains a high R_{ct}, this solder exhibits a tendency for extremely slow migration growth.

On the other hand, although Sn–37Pb forms passivity film layers with materials such as SnO_2 and PbO, these passivity films have poor stability, and so the R_{ct} drops rapidly and this type of solder exhibits a tendency for faster migration growth.

Because lead-free solder forms a more stable passivity film than conventional Sn–Pb eutectic solder, the charge-transfer resistance of lead-free solder is higher and more stable. Because of this, lead-free solder is presumed to suppress the growth of migration because it suppresses charge transfer and material transfer.

7.5.2 Analysis of the Cathode Deposition Process Using Water Drop Test (13)

A water drop test (using such standards as IPC-TM650-2.6.13) was performed to analyze cathode deposits. The electrodes were comb pattern (IPC-

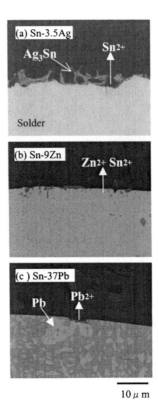

Figure 7.11   Anode cross-section observation following migration testing. (From Ref. 16.)

B-25 test board) with 0.318-mm spacing, and the electrode surfaces were plated with solder. One microliter of ion exchange water was dripped onto the interval between one pair of electrodes, 5 V/d.c. was impressed, and migration occurred at room temperature.

Shorting between electrodes caused by migration occurred at the following times: Sn–37Pb at about 10 sec, Sn–3.5Ag at about 120 sec, and Sn–9Zn at about 330 sec. Lead-free solder showed an overall tendency toward comparatively longer times before the occurrence of migration than conventional leaded solder (Sn–37Pb).

Figure 7.13 shows SEM image of migration deposits after the water drop test. Sn–3.5Ag exhibited needle-shaped migration extending from the cathode toward the anode. The deposits were composed of Sn. Sn–9Zn exhibited dendritic migration, widely branched, growing from the cathode

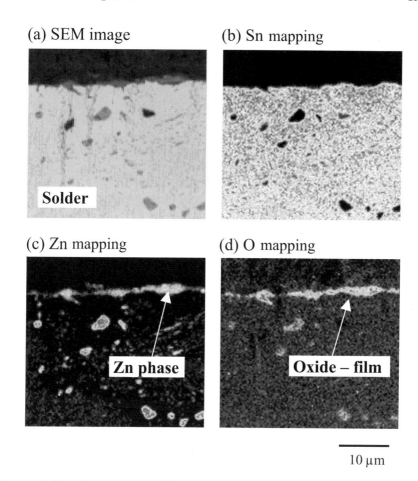

(a) SEM image   (b) Sn mapping

Solder

(c) Zn mapping   (d) O mapping

Zn phase   Oxide – film

10 μm

Figure 7.12 Cross-sectional EPMA analysis of the anode electrode of Sn–9Zn solder after the migration testing.

toward the anode. The migration was composed of Zn and Sn. On the other hand, Sn–37Pb experienced multiple-branching migration composed of Pb and Sn and covering the electrode gaps. This migration material is thought to come from Sn and Pb dissolving and being deposited.

These results show that cathode migration deposits are mainly composed of the same elements that dissolve from the anode, and the solder alloy migration characteristics are affected by the anode dissolution characteristics.

Table 7.4 shows the relationships of the solder alloy surface structures, the surface passive films, and the cathode deposits for the above-mentioned types of solder.

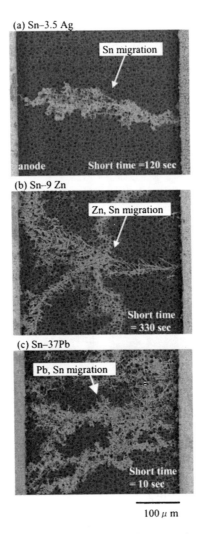

(a) Sn–3.5 Ag

Sn migration

anode   Short time =120 sec

(b) Sn–9 Zn

Zn, Sn migration

Short time = 330 sec

(c) Sn–37Pb

Pb, Sn migration

Short time = 10 sec

100 μ m

Figure 7.13   Photographs of migration after the water drop test. (From Ref. 13.)

7.6   RELIABILITY TESTING

This section will present the evaluation results of the temperature–humidity–bias test and evaluation of surface insulation for Sn–Ag–Cu solder, which is now being used in commercial products.

Table 7.4 Solder Surface Structure and Migration Characteristics

Solder composition	Sn–Ag	Sn–Cu	Sn–Bi	Sn–Zn	Sn–Pb
Surface structure	Sn;	Sn;	Sn;	Sn;	Sn;
	Ag$_3$Sn	Cu$_6$Sn$_5$	Bi	Zn	Pb
Surface passive film	SnO$_2$	SnO$_2$	SnO$_2$	ZnO; Zn(OH)$_2$; SnO$_2$	Pb$_x$O$_x$; SnO$_2$
Metal deposits	Sn	Sn	Sn	Zn, Sn	Pb, Sn

7.6.1 Surface Insulation Resistance Test (14)

For insulation resistance tests, we used the solder alloy Sn–3.5Ag–0.75Cu, which is widely used in market applications. Test specimens were used comb-pattern (IPC-B-25 test board) electrodes with 0.318-mm spacing on a PWB (NEMA FR-4) using rosin activity flux and hot dipping. The test was a temperature–humidity–bias test (using such standards as ISO 9455-17) at 85°C, 85%, and 50 V/d.c. bias voltage. During the test, insulation resistance values were recorded at constant intervals. Figure 7.14 shows changes in resistance values during the test. Insulation resistance was maintained at approximately 1×10^{10} Ω throughout the test, and no insulation degradation was seen at 2000 hr. External observation after the test also revealed no migration growth.

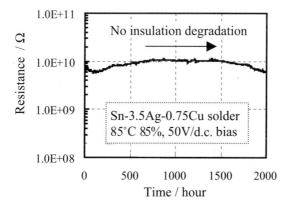

Figure 7.14 Change in insulation resistance of Sn–Ag–Cu solder during THB test. (From Ref. 14.)

7.6.2 Evaluation Using Mounting PWBs

Reliability was evaluated using PWBs with surface-mounted devices. The test specimens were prepared by reflow soldering with Sn–Ag–Cu solder paste. The test was a temperature–humidity–bias test (80°C, 90%, 5 V/d.c. bias voltage), with specimens removed after 1000 hr to check operation and perform visual inspection. Figure 7.15 shows a photo of a quad flat package

(a)   Terminals of QFP

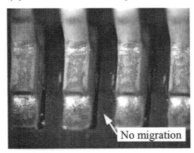

(b)   Capacitor

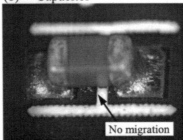

(c)   Resistance

0.5mm

Figure 7.15 External appearance of Sn–Ag–Cu solder joints after the THB test (80°C/90%, 1000 hr, 5 V/d.c. bias).

terminal interval (0.5-mm spacing), a chip capacitor (1608 type, with 0.8-mm spacing), and a chip resistance (2025 type, with 1.2-mm spacing). After 1000 hr of testing, the specimens showed no evidence of function degradation nor of migration growth.

The results of the above two reliability tests do not confirm the occurrence of migration or insulation degradation in lead-free solder even for components mounted on PWBs. Therefore, it is believed that lead-free solder may have the capability of maintaining electrical reliability.

7.7 SUMMARY

This chapter has looked into the relationship between solder surface structure and anode dissolution characteristics and discussed the factors involved in the occurrence of migration in lead-free solder through an analysis of the migration process based on electrochemical techniques. Long-term electrical reliability has been considered according to reliability tests.

The indicate that lead-free solder with Sn as its main component forms a stable passivity film on the surface of the electrodes and so maintains a higher electrical reliability than conventional Sn–Pb eutectic solder.

Ag and Cu, which have conventionally exhibited fast migration occurrence, have been shown to form stable compounds that do not dissolve in solution. With Sn–9Zn, the base metal Zn dissolves out before Sn, but in that process, a passivity film is formed, and so plays the role of inhibiting further dissolution of Zn.

However, these passivity films may have their stability damaged by environmental conditions. Because of this, the usage environment must be carefully considered.

REFERENCES

1. G Digiacomo. Metal migration (Ag, Cu, Pb) in encapsulated modules and time-to-fail model as a function of the environment and package properties. 20th IEEE/PROC. NJ: IEEE International Reliability Physics Symposium, San Diego, CA, 1982, pp 27–33.
2. PH Dumolin, JP Seurin, P Marce. Metal migration outside the package during accelerated life tests. IEEE trans components, hybrids, manufacturing tech, CHMT-5, 479–486, 1982.
3. JJ Steppan, JA Roth. A revue of corrosion failure mechanisms during accelerated tests—electrolytic metal migration. J Electrochem Soc 134:175–190, 1987.
4. H Tanaka, F Ueta. Effects of reflow processing and flux residue on ionic migration of lead-free solder plating using the quartz crystal microbalance method. Mater Trans 42:2003–2007, 2001.

5. H Tanaka, S Yoshihara. Investigation on ionic migration phenomenon of lead-free solder. Kagaku-to-Kogyo 76:503–512, 2002.

6. IPC electrical task group: electrochemical migration: electrically induced failures in printed wiring assemblie, Standard-IPC-TR-476A. IL: IPC, 1997.

7. IPC surface insulation resistance task group: surface insulation resistance handbook, Standard-IPC-9201. IL: IPC, 1996.

8. GT Kohman, HW Hermance, GH Downes. Silver migration in electrical insulation. Bell Syst Tech J 34:1115–1147, 1955.

9. SJ Krumbein. Electrolytic model for metallic electromigration failure mechanism. IEEE Trans Reliab 44:539–549, 1995.

10. K Sauter. Electrochemical migration testing results—evaluating PWB design, manufacturing process, and laminate material impacts on CAF resistance. IPC Revue, August, 8–18, 2002.

11. K Tsuruta, S Yoshihara, T Shirakashi. Investigation of migration in printed circuit boards. J Surf Finish Soc Jpn, 48:84–88, 1997.

12. M Pourbaix. Atlas of electrochemical equilibria in aqueous solutions. Houston: NACE, 1966.

13. H Tanaka. Factors leading to ionic migration in lead-free solder. ESPEC Tech Rep 14:1–6, 2002.

14. H Tanaka, Y Saito. Investigation into the electrochemical migration process of solder alloys. IPC/PROC. IPC Printed Circuits Expo IL:IPC, 2003, Long Beach, CA, session 1t-2-2.

15. H Tanaka, H Hiramatsu. Analysis on initiation process of ionic migration for lead-free solder by use of AC impedance method. J Jpn Inst Electron Packag 5:188–191, 2002.

16. H Tanaka, M Yamashita. Investigation on ionic migration phenomenon of hot dipped lead-free solders using QCM method. J Jpn Inst Electron Packag 2002; 5:135–139.

8

Lead-Free Solder Paste and Reflow Soldering

Shinichi Fujiuchi

Sanmina-SCI Systems Japan, Ltd., Shiga-ken, Japan

8.1 INTRODUCTION

In automated soldering of a printed wiring board (PWB), there are two basic methods: wave soldering and reflow soldering. The wave soldering method applies molten solder to the bottom side of a PWB. Application of solder and heating are done simultaneously in the wave soldering. In reflow soldering, however, solder paste is printed on a PWB before heating. A controlled amount of solder and flux is applied to the area to be soldered, which results in higher assembly yield. A common technique of applying solder and flux is to print a pattern of solder paste. Solder paste becomes an indispensable material for the electronics assembly process.

Solder paste is a homogeneous mixture of solder powder, flux, and vehicle. It is applied on Cu pads or component terminals to be soldered with screen printing or dispensing process and can form a metallurgical bonding at a heating process. In the electronics industry, solder paste has a minor role in production volume compared with solder bars, but it plays an essential role as an interconnecting material in the surface mount technology (SMT) process. During the last two decades, solder paste has shown a prodigious progress in its performance, including printability of fine pitch pattern, rheological stability, and solderability. This has been done to meet market requirements toward high-density packaging. It allows for better soldering quality and high-density packaging when used for, for instance, 0.4-mm QFP (Quad Flat Package) and even flip chip assembly on organic carriers [1]. However, in many cases, it has been developed on an assumption of the usage of Sn–Pb eutectic solder.

Sn–Pb eutectic solder has advantages in physical and chemical properties. Its lower melting point minimizes the thermal damage of components during the soldering process. Its good wettability allows lower soldering defect rate, and low yield point easily releases an internal stress at a soldered joint. The soldering parameters, such as printing pattern and reflow heating profiling have been optimized based on these favorable properties of Sn–Pb eutectic solder.

In recent years, the usage of Pb, which can be harmful to the human body, has attracted much global attention. The European Union (EU) is making an effort to regulate the usage of Pb, and according to the EU directive on WEEE/RoHS that is mentioned in Chapter 1 in detail, electrical appliances must be lead free starting in 2006.

A lead-free product requires lead-free solder, components, and raw boards. In the case of solder paste, the solder composition change impacts process parameters, such as reflow temperature profiling, reflow atmosphere, and paste printed patterns, due to differences of physical and chemical properties between Sn–Pb eutectic and lead-free solders.

This chapter is intended to cover general descriptions and requirements of solder paste, key physical and chemical properties of lead-free solder paste, and to illustrate evaluation methods of physical properties of solder paste. The outline of the reflow process is also introduced.

8.2 GENERAL DESCRIPTION OF SOLDER PASTE

8.2.1 Reflow Soldering Process and Key Requirements of Solder Paste

An example of the reflow soldering process for single-sided board is shown in Fig. 8.1. For double-sided board, this procedure has to be performed for each side. The assembly procedure consists of five steps: Step 1, paste application; Step 2, component placement; Step 3, preheating; Step 4, reflow; and Step 5, cooling. Key requirements of solder paste for each step are described as follows:

Step 1: Paste Application   Solder paste is transferred onto a substrate or component terminals with various methods of metal stencil printing, mesh screen printing, dispensing, pin transfer, and so on. The amount and location of solder paste on a substrate are determined at this step.

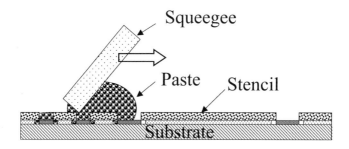

Step 1 Paste apply

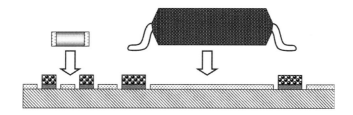

Step 2 Component placement

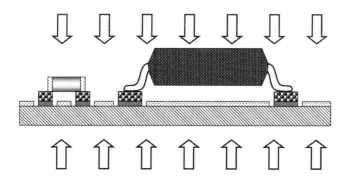

Step 3 Preheating

Figure 8.1 Assembly procedure with solder paste.

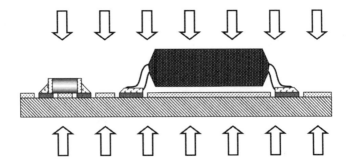

Step 4 Reflow

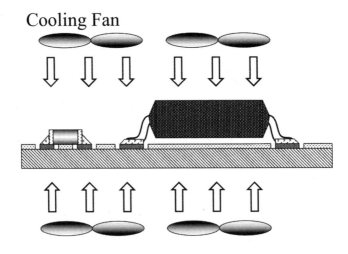

Step 5 Cooling

Figure 8.1 Continued.

Key requirements of solder paste at this step are as follows:

- Printability of paste at the printing process.
- Dispensability through a needle at the dispensing process.
- Chemical and physical stability (shelf life and exposure time).
- Small cold slump.

Step 2: Component Placement Components are placed on a substrate. Component terminals make contact with the solder paste and the tackiness of the solder paste prevents misalignment of the components before soldering. Paste patterns are deformed by compression of component terminals, which may result in a soldering defect after reflow soldering.

Key requirements of solder paste at Step 2 are as follows:

- Tacking force/time.
- Small cold slump at/after component placement.

Step 3: Preheating After placement of components, a substrate is heated at a specified temperature below the melting point of the solder.

Purposes of preheating are the following:

- Homogeneous temperature distribution in a substrate.
- Prevention of thermal shock of components.
- Prevention of solder splash because of bumping of organic solvents.

In this step, the solid content of flux is gradually increased due to evaporation of the organic solvent. Viscosity of flux will be affected by hardening of flux due to high solid content and softening of flux at preheating temperature.

Deformation of the paste pattern sometimes causes a soldering defect after reflow soldering. Therefore, key requirement of solder paste at Step 3 is as follows:

- Small hot slump.

Step 4: Reflow After preheating, solder paste is heated at a temperature above the melting point of the solder. The solder is melted at Step 4, and wet on the metal surface to be soldered.

Key requirement of solder paste at Step 4 is as follows:

- Solderability

Step 5: Cooling Molten solder is cooled and solidified during Step 5. After solidification, the soldered joints show mechanical strength.

After the cooling process of Step 5, flux residue is sometimes removed to ensure long-term reliability. In this case, flux residue has to have a good cleanability. In a case of no cleaning, flux residue has to have low corrosivity, low electromigration, and high electric resistivity.

8.3 UNIQUE REQUIREMENTS OF LEAD-FREE REFLOW SOLDERING

There are some unique requirements for lead-free solder paste because the physical and chemical properties of lead-free solders are different from those of Sn–Pb eutectic solder.

8.3.1 Heat Resistivity

Lead-free solder requires higher reflow temperature because the melting points of many lead-free solders are higher than those of Sn–Pb eutectic solder.

Many lead-free solders were proposed and some of them have already been patented in the market. Basic compositions of lead-free solder are Sn–Ag, Sn–Cu, and Sn–Bi for binary alloys, and Sn–Ag–Cu and Sn–Zn–Bi for ternary alloys. Compositions and melting points of some basic lead-free solders are shown in Table 8.1. Lead-free solders except the Sn–Bi solder shown in the table have higher melting points than Sn–Pb eutectic solders (183°C).

In board assembly with solder paste, Sn–Ag–Cu solder is considered the leading candidate of lead-free solders because it produces highly reliable solder joints. Sn–Cu solder is not considered suitable for reflow soldering

Table 8.1 Composition and Melting Point of Lead-Free Solder

| | Solder composition | | | | | | | Solidus temperature (°C) | Liquidus temperature (°C) |
	Sn	Ag	Cu	Sb	Zn	Bi	Pb		
Lead-free	95			5				240	243
	99.25		0.75					227	227
	96.5	3.5						221	221
	95.75	3.5	0.75					217	219
	95.5	3.9	0.6					217	217
	96.5	3	0.5					217	220
	91				9			199	199
	89				8	3		190	197
	42					58		139	139
Ref.	63						37	183	183

because of its higher melting point and lower reliability compared with Sn–Ag–Cu solder. It is mainly utilized in wave soldering process due to its low material cost. Sn–Bi solder has a low melting point (139°C), which is an advantage in preventing thermal damage of components in the soldering process. The Sn–Bi solder, however, has a disadvantage in its mechanical properties. Although it was reported that addition of Ag in Sn–Bi solder greatly improved its mechanical properties, further study will be required. Sn–Zn and Sn–Zn–Bi solders have intermediate melting temperatures compared with Sn–Ag–Cu and Sn–Pb solders, and they have an advantage in preventing thermal damage in reflow soldering. However, further studies are required to ensure long-term reliability because Zn is an easily oxidized and corrosion-prone metal and combination effect on joint reliability with a terminal plating was reported [2].

Figure 8.2 illustrates the thermal conditions of reflow soldering with Sn–Pb eutectic solder and Sn–Ag–Cu solder, the leading candidate of lead-free solders for reflow soldering. In this figure, reflow peak temperature is assumed to be more than 205°C for Sn–Pb eutectic solder and 225°C for Sn–Ag–Cu lead-free solder. Maximum peak temperature of the component body is assumed to be less than 240°C. Based on the assumption described earlier, the process window of reflow peak temperature is 35°C (= 240–205) for Sn–Pb solder and 15°C (240–225) for Sn–Ag–Cu solder. In an actual board assembly, many components that have various thermal capacities are mounted on the same board, and the reflow temperatures of the components deviate from each other. In the case of Sn–Pb solder, a temperature deviation of 35°C is acceptable because its process window is wide. On the other hand, a temperature deviation less than 15°C is required due to the narrow process window of lead-free soldering.

To minimize the temperature deviation, higher preheating temperature and longer preheating duration are required. These thermal conditions, such as higher reflow temperature, higher preheating temperature, and longer preheating duration, tend to oxidize the solder surface and degrade fluxes.

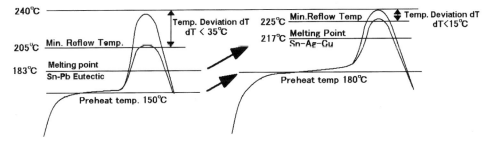

Figure 8.2 Comparison of thermal conditions for Sn–Pb and Sn–Ag–Cu solder.

Figure 8.3 shows the influence of preheating conditions, such as temperature, duration, and atmosphere, on solderability [3].

Solder paste that was air-reflowed showed smooth surfaces for preheating conditions of 150°C/120 sec, 175°C/60 sec, and 200°C/40 sec, but odd-shaped solder surfaces for 200°C/60 sec. Solder paste that was air-reflowed under preheating conditions of 150°C/180 sec and 175°C/90 sec take a mean surface condition between the two.

With nitrogen reflow, the solder surface obtained was smooth even with a preheating condition of 200°C/180 sec.

This experimental result showed the importance of preheating conditions on the solderability of solder paste. Fig. 8.4 shows an example of a nitrogen reflow furnace for lead-free soldering.

To improve heat resistivity, modified rosin flux is used for lead-free solder paste.

Rosin is a true resin that originates from pine trees; it is characterized as gum rosin, wood rosin, and tall oil rosin. The major components of rosin are abietic acid, neoabietic acid, isopimaric acid, and pimaric acid. These components belong to a carboxyl group, and have condensed three-ring structure, and a double bond or a conjugated double bond. They can be represented by

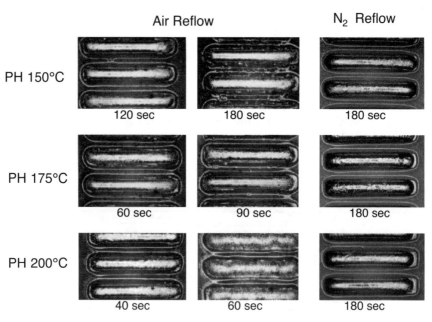

Figure 8.3 Influence of preheating conditions on solderability of lead-free solder paste.

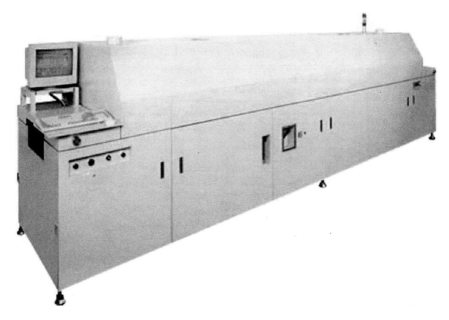

Figure 8.4 Nitrogen reflow furnace for lead-free soldering. (Courtesy of Senju Metal Industry Co., Ltd.)

the formula $C_{19}H_{28}COOH$ or $C_{19}H_{31}COOH$. True resins are easily oxidized and recrystallized, which can result in an increase of their melting points and degradation of their properties. The stability of true rosin is modified through organic disproportionation, hydrogenation, polymerization, esterification, and so on. Figure 8.5 shows examples of modification processes for abietic acid.

8.3.2 Hot Slump Due to Higher Preheating Temperature

As mentioned in Section 8.3.1, the preheating temperature of lead-free soldering is usually higher than that of current Sn–Pb eutectic solder, which may result in larger hot slump and a deformation of paste pattern during preheating. Fig. 8.6 shows the differences of hot slump among three types of solder pastes. Paste A showed practically no hot slump, and paste C showed large hot slump. Hot slump is one of the causes of solder bridges and solder balling. Figure 8.7 shows solder balling beside a capacitor chip after reflow soldering, which was generated by deformation of solder paste.

The hot slump is affected by the viscosity of the flux, which is varied by change in solid content due to evaporation of organic solvent and softening of

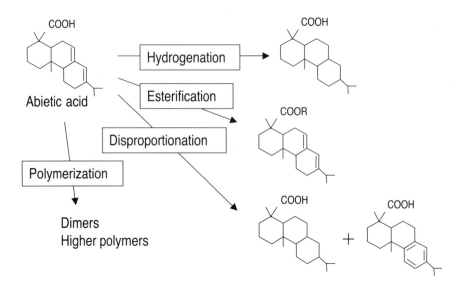

Figure 8.5 Modification of abietic acid.

flux during preheating. It is important to evaluate the hot slump in developing and selecting a lead-free solder paste. The evaluation method is discussed in Section 8.4.5.

8.3.3 Poor Spreading of Lead-Free Solder

Solder spreading of lead-free solder is poor and the paste printing pattern may have to be readjusted.

As shown in Table 8.1, most lead-free solders have high Sn content. Figure 8.8 shows the effect of Sn content on surface tension of molten Sn–Pb alloy at $350°C$ under inert atmosphere [4]. Higher Sn content shows higher surface tension. From Young's equation, a solder whose surface tension is

Paste A   Paste B   Paste C

Figure 8.6 Hot slump during preheating.

Figure 8.7 Solder ball beside a capacitor chip.

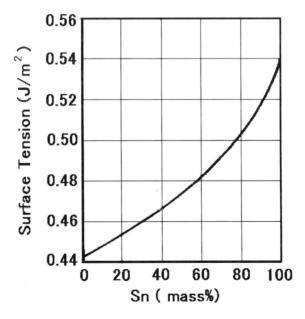

Figure 8.8 Effect of Sn content on surface tension of Sn–Pb molten solder.

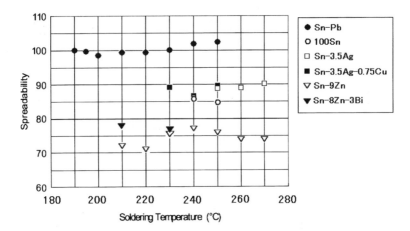

Figure 8.9 Spreadability of Sn base alloy on Cu plate (spreadability of Sn–Pb at 190 °C = 100).

high shows higher contact angle, therefore lower solder spreading. Figure 8.9 shows solder spreading of some lead-free solders and Sn–Pb eutectic solder [5]. The spreadability in this figure is an index of solder spreading compared with the spreading of Sn–Pb solder at 190 °C (= 100%). Sn–Ag–(Cu) and Sn–Zn–(Bi) show spreadability of 85–90% and 70–80%, respectively. The spreading of lead-free solders is almost stable even in elevated reflow temperature. Spreading of Sn–3Ag–0.5Cu lead-free solder and Sn–Pb eutectic solder on QFP pads are shown in Fig. 8.10. Whole Cu surfaces were covered with Sn–Pb solder, but exposed Cu surface was found when using Sn–Ag–Cu solder.

In current SMT assembly process with Sn–Pb eutectic solder, stencil openings often shrink in size to control solder amount and reduce soldering

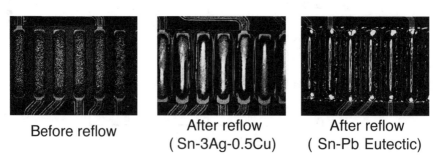

Figure 8.10 Comparison of solder spreading on Cu pads.

defects, for example, solder short and solder balling. If whole Cu surfaces have to be covered with lead-free solder, the stencil opening cannot shrink in size, which may result in soldering bridge and solder balling because of the large amount of solder. If better solder spreading is required in an application, surface finishing of the substrate, such as solder coating or plating, has to be considered.

8.3.4 Chemical Reaction Between Lead-Free Solder and Flux

Solder paste is a homogeneous mixture of solder powder, flux, and vehicle. In general cases, fine and spherical particles are used for solder paste to improve its printability for fine pitch pattern. Figure 8.11 shows SEM micrographs of solder particles whose solder compositions are Sn–3Ag–0.5Cu, Sn–9Zn, and Sn–Pb eutectic. The solder particles suspended in solder paste make contact with soldering flux, which has a corrosive action in the soldering process. The flux is designed to be active at soldering temperatures and inactive at room temperature. The corrosive action, however, still remains at room temperature and corrodes the solder surface. The reaction is described by the following equations.

$$MO + 2R\text{-}CHOOH \; \rightarrow \; (R\text{-}COO)_2M + H_2O \tag{8.1}$$

$$M + 2R\text{-}COOH \; \rightarrow \; (R\text{-}COO)_2M + H_2 \tag{8.2}$$

$$MO + 2R\text{-}NH_2HBr \; \rightarrow \; MBr_2 + 2R\text{-}NH_2 + H_2O \tag{8.3}$$

$$M + 2R\text{-}NH_2HBr \; \rightarrow \; MBr_2 + 2R\text{-}NH_2 + H_2 \tag{8.4}$$

These equations show that activated flux, solder composition whose ionization tendency is high, oxidized solder, and fine particles that have large specific surface accelerate a corrosion action between solder and flux.

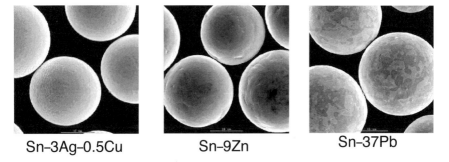

Sn–3Ag–0.5Cu   Sn–9Zn   Sn–37Pb

Figure 8.11 SEM micrographs of solder particles. (Courtesy of Nihon Genma Mfg. Co., Ltd.)

 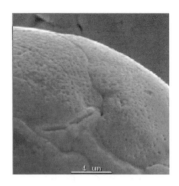

(a) Before contact with flux   (b) After contact with flux

Figure 8.12   Surface of Sn–8Zn–3Bi solder particle. (Courtesy of Nihon Genma Mfg. Co., Ltd.)

This reaction promotes recrystallization of rosin and degradation of flux properties.

It is well known that Sn–Zn and Sn–Zn–Bi pastes show rapid increase in viscosity and degradation of solderability. A recent study shows selective corrosion of Zn on solder particles (see Fig. 8.12) and the effectiveness of organic coating to prevent corrosion reaction between solder and flux [6].

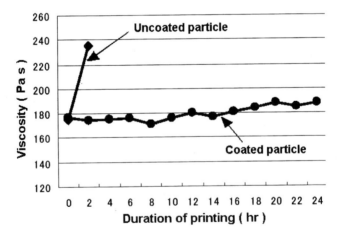

Figure 8.13   Viscosity versus time to read in continuous printing of Sn–8Zn–3Bi lead-free solder paste.

Figure 8.13 shows viscosity versus time to read in continuous printing for solder pastes with coated and noncoated Sn–8Zn–3Bi solder particles. Solder paste with coated particles shows stable viscosity.

8.4 EVALUATION METHOD OF SOLDER PASTE

Virtually every document that specifies test procedures of solder pastes is mainly for Sn–Pb eutectic solder. However, many of them are applicable to lead-free solder paste because the specified test conditions link to the melting points of solder. Some test procedures for evaluating lead-free solders are introduced in the following.

8.4.1 Tackiness

Tackiness is one of the advantages of solder paste, which enables components to be held before soldering. Solder paste just after printing on a substrate usually has sufficient tackiness for a length of time, but it will be reduced during exposure to ambient atmosphere.

Tackiness of solder paste is one of the requirements for solder paste in surface mount technology and is expressed in tacking force and tacking time. Tacking force is defined as the peak force required to maintain contact between a component and solder paste. Tacking time is the length of time in which the tacking force is kept over a specific value. They depend on component shape and surface conditions of a component and solder paste. In the case of solder paste, tacking force is influenced by flux content in solder paste, flux composition, condensation of solid flux on paste surface due to evaporation of solvent, and so on. As mentioned in Section 8.3, some properties of lead-free solder pastes are different from those of Sn–Pb solder paste and some constituents in solder paste are replaced to meet the unique requirements of lead-free soldering. For example, use of a solvent that has higher vapor pressure to prevent hot slump may affect the short tacking time of solder paste. Solder paste should be designed to have proper tacking force and time for an application.

Tacking test procedure is proposed in the following three documents:

J-STD-005/IPC-TM-650 2.4.44.
IIW SC/1A-SP058/87.
JIS Z 3284 Annex 9.

In these test procedures, tacking force is measured with a stainless steel probe. The probe brings to a solder paste patterns that were printed on a substrate with a stencil, and makes contact with the solder paste under a

Table 8.2 Test Conditions for Tackiness Measurement

		JIS Z3284 Annex 9	ANSI/ IPC-SP-819	IIWDuc.SC/ 1ASP058/87
Stencil	Diameter	6.5	6.35 (0.25 in.)	5
	Thickness	0.2	0.254 (0.01 in.)	0.2
Probe	Diameter	5.1 ± 0.13	5.08 ± 0.13 (0.20 in.)	1.6 ± 0.05
Test condition	Probe speed, downward (mm/sec)	2	2.54 ± 0.5	30 ± 2
	Pressure	50 ± 5 g	300 ± 30	30 ± 0.2
	Bench time after pressurization	0.2 s	5	5 ± 1
	Probe speed, upward (mm/sec)	10	2.54 ± 0.5	30 ± 2

specified condition. After a specific bench time, the probe is lifted under specified condition to measure the tacking force, which is the peak force during the lifting. Table 8.2 shows a comparison of test conditions among J-STD, IIW, and JIS standards. Figure 8.14 illustrates the sequence of probe motion and force added to the probe. A tacking test equipment used for solder paste is shown in Fig. 8.15, which can be applicable to the tacking test specified by IPC, IIW, and JIS standards.

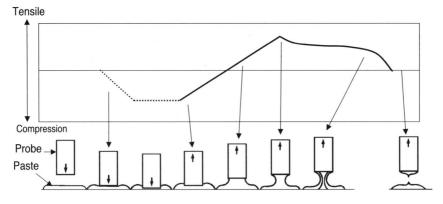

Figure 8.14 Sequence of sample motion and force curve.

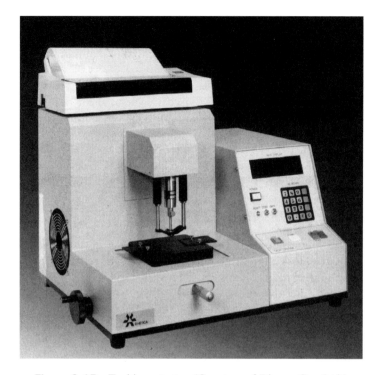

Figure 8.15 Tackiness tester. (Courtesy of Rhesca Co., Ltd.)

A simple procedure for qualitative evaluation of the tackiness is described as follows:

1. Print a specified pattern (e.g., QFP, BGA [ball grid array], etc.) with solder paste on a substrate.
2. Place the component after a specified bench time (e.g., 30 min, 1 hr, 2 hr, 5 hr).
3. Turn the substrate upside down.
4. Measure the time when the component drops.

This simple test method is effective in making a comparative study of soldering parameters, such as solder pastes, paste pattern, and exposure time.

8.4.2 Viscosity

Viscosity is one of the most important parameters of solder paste. Printing performance strongly depends on the viscosity. Thick solder paste, whose viscosity is high, sometimes results in insufficient printed pattern or, occasionally, lack of paste on pad, which results in poor solder and non solder after reflow soldering. Thin solder paste, whose viscosity is low, may lead to cold or hot slumping and soldering defects such as short and solder balling after reflow soldering. Viscosity control is essential to maintain high productivity in a board assembly process.

In measuring the viscosity of solder paste, there are some basic parameters to be measured, such as "viscosity versus shear rate," "viscosity versus time," and "yield point." They are obtained by adding various shear rate to solder paste. Figure 8.16 shows three examples of viscosity sensors available to measure the viscosity of solder paste. These sensors have simple shapes and have an advantage in making a model for calculating the viscosity of solder paste. However, in practical operation for measuring the viscosity of solder paste, thick solder paste, whose viscosity is high, sometimes makes a gap between the sensors and provides erroneous data.

Figure 8.17 shows a spiral sensor. This viscometer utilizes a double-cylinder spiral pump method. The inner cylinder has spiral flutes making it resemble a screw. It is attached to a torque sensor. The outer cylinder has an inlet scoop and exhaust openings. The outer cylinder can rotate at a prescribed speed. Because of the geometry of the two cylinders, the assembly is a type of pump. In the measuring operation, the paste is scooped into the gap between the cylinders and sheared between the inner and outer cylinders. Key parameters of dynamic viscosity can be measured with the relationship between torque, rotation speed, and duration time. Figure 8.18 shows a spiral-type viscometer.

J-STD-005/IPC-TM-650 2.4.34, 2.4.34.1-3, and JIS Z 3284 Annex 6 specify the testing procedures.

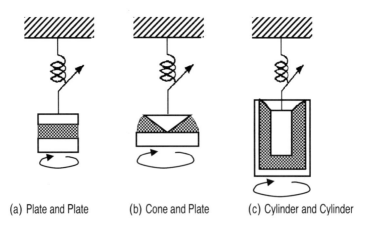

(a) Plate and Plate   (b) Cone and Plate   (c) Cylinder and Cylinder

Figure 8.16 Schematic of three types of viscosity sensors.

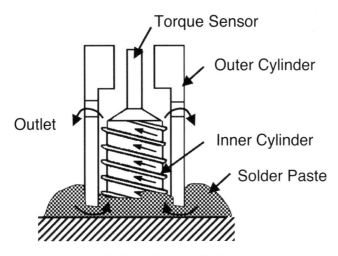

Figure 8.17 Schematic of spiral sensor.

8.4.3 Solder Balling

Solder balling test is carried out to evaluate the ability of coalition of solder particles in solder paste. JIS Z 3284 Annex 11 and J-STD-005/IPC-TM-650 Method 2.4.43 specify the test procedure.

1. Print a solder paste on nonwettable substrate, such as Al_2O_3 substrate or a frost glass, with a stencil.
   The opening size of the stencil is 6.5 mm in diameter. Thickness of the stencil is specified as 0.2 mm in JIS Z3284 and 0.15 and/or 0.2 mm in J-STD-005.
2. Check the diameter and thickness of printed paste.
3. Place the substrate on the surface of molten solder and dissolve the solder paste after a specified storage condition.
   Storage conditions for each specification are as follows:

   JIS Z 3284

   a. Within 1 hr after printing.
   b. After 24-hr storage under conditions of relative humidity of 60 ± 20% and temperature of 25 ± 2°C.

   J-STD-005

   a. Within 15 ± 5 min after printing.
   b. After 4-hr storage under relative humidity of 50 ±10% and temperature of 25 ± 3°C.

Figure 8.18   Spiral viscometer. (Courtesy of Malcom Co., Ltd.)

The temperature of molten solder is the liquidus temperature of the alloy plus 50°C.

4.   After cooling, particle size and number of the solder balls remaining around the coalesced solder are observed by using a magnifier. Evaluation is made according to the criteria and photograph of boundary samples.

   JIS Z 3284 specifies four degrees based on ball size and number of balls.

Aggregation degree 1: Solder pastes were melted to make a large coalesced solder ball. No solder ball is found around it.

Aggregation degree 2: Solder pastes were melted to make a large coalesced solder ball. Three or less solder balls whose diameter is 75 μm or under are observed around it.

Aggregation degree 3: Solder pastes were melted to make a large coalesced solder ball. Four or more solder balls whose diameter is 75 μm or under are observed around it, but the balls do not form a semicontinuous annular form.

Aggregation degree 4: Solder pastes were melted to make a large coalesced solder ball. A number of solder balls form a semicontinuous annular form around the large coalesced solder ball.

8.4.4 Solderability Test

Solderability of solder paste is the ability to wet the surface to be joined with coalescence of solder particles, and make the joints with smooth solder surface. Spreading of reflow-soldered paste or wetting force in soldering process is measured for the evaluation.

JIS Z 3284 Annex 10 and J-STD-005/IPC-TM-650 Method 2.4.45 provide qualitative methods for the evaluation of the solderability of solder paste. In these methods, the solderalibity is estimated based on the spreading of solder after heating, smoothness of solder surface, i.e., no evidence of dewetting or nonwetting, and no solder spatter around the printed dots. The solderability is subdivided into four classes in JIS Z 3284.

Class 1: Solder paste is dissolved. The wetting area becomes larger than the area that is paste coated.

Class 2: The entire area coated by solder paste is covered by dissolved solder.

Class 3: Almost all the area covered by solder paste is wetted by the molten solder (including dewetting area).

Class 4: No solder seems to wet, and molten solder becomes one or more solder balls (nonwetting).

Quantitative evaluation method to determine the solderability of component terminals with solder paste is specified in Japan Electronics and Information Technology Industries Association (JEITA) ET-7404. In this method, a meniscograph, which has a load sensor connected to a sample, provides information on the solderability by using a wetting balance to monitor the wetting force in a solder paste or molten solder as a function of time. In JEITA ET-74404, two types of heating procedures are specified: (1) gradual heating with preheating and (2) rapid heating without preheating. Figure 8.19 illustrates schematics of solderability testers for gradual and rapid heating. The solderability tester of gradual heating has an electric heater that can heat solder paste in accordance with a programmed temperature profile. The solderability tester of rapid heating has a solder bath to heat solder paste.

A test procedure of gradual heating is described as follows (Fig. 8.20).

1. Dip the specimen slightly into solder paste at t_0.
2. Start heating by electric heater at t_1.
3. Hold for a specific duration.
4. Extract the specimen from molten solder at t_5.

A test procedure of rapid heating is described as follows (Fig. 8.21).

1. Dip the specimen slightly into solder paste at t_0.
2. Move a solder pot upward to make a contact between molten solder and paste plate.
3. Make a contact and start heating at t_1.
4. Hold for a specific duration.
5. Extract the specimen from molten solder at t_5.

When the specimen is immersed in solder paste at t_0, upward force is exerted on it. Softening of solder paste, melting, and wetting of solder occurs between t_1 and t_2. At t_2, upward buoyancy and downward wetting forces are

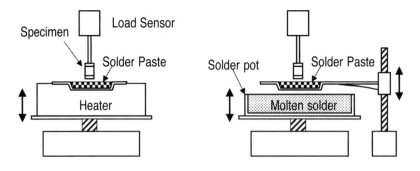

(a) Step heating with preheating   (b) Rapid heating without preheating

Figure 8.19   Schematic of solderability tester with solder paste.

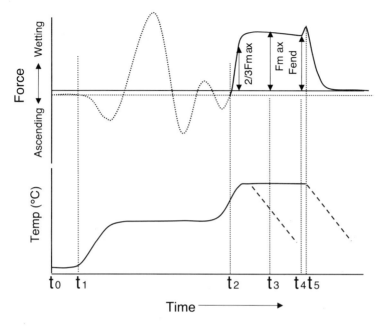

Figure 8.20 Temperature profile and force curve (gradual heating).

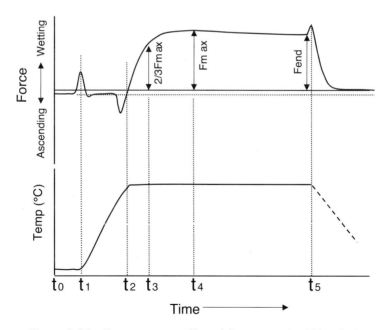

Figure 8.21 Temperature profile and force curve (rapid heating).

balanced, which results in zero net force. Solderability is expressed as incubation period of wetting (T_0), wetting time (T_1), and stability of wetting (S_b), where

$$T_0 = t_2 - t_1 \quad \text{(for rapid heating)} \tag{8.5}$$

$$T_0 = t_3 - t_2 \quad \text{(for gradual and rapid heating)} \tag{8.6}$$

$$F_b = F_{\text{end}}/F_{\text{max}} \quad \text{(for gradual and rapid heating)} \tag{8.7}$$

where T_3 is the time that wetting force reached two thirds of F_{max}, and F_{max} is the maximum wetting force between t_1 and t_5.

T_0 is a transition phase where solder particles in solder paste coalesce into large units. Net force at T_0 is a combination of the wetting force of solder paste and coalesced molten solder, and buoyancy of specimen in solder paste and coalesced molten solder. JEITA ET-7404 does not define T_0 in gradual heating and net force at T_0 in gradual and rapid heating because further study is required to clarify the physical meaning of a force at T_0.

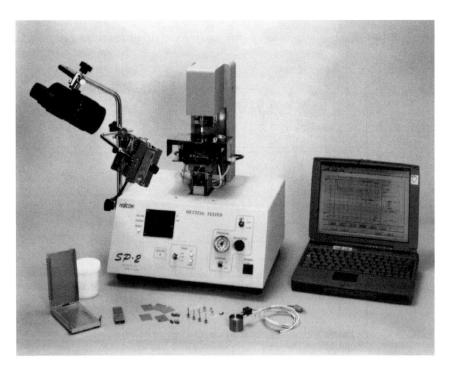

Figure 8.22   Solderability tester SP-2. (Courtesy of Malcom Co., Ltd.)

There are no universal values of T_0, T_1, and F_b to be followed because these parameters strongly depend on component shape, size, and heat capacity. These parameters also depend on the melting point and surface conditions of solder and surface finish of terminals. In lead-free soldering, T_0 and T_1 trend to increase due to their higher melting points. The criteria for solderability with lead-free solder should be established under a new set of assembly conditions.

Figure 8.22 shows a solderability tester for paste and components.

8.4.5 Slump Test

Slump test is carried out to evaluate the ability of keeping printed pattern of solder paste. JIS Z 3284 Annex 7 and 8 and J-STD-005/IPC-TM-650 Method 2.4.35 specify the test procedures.

1. Print a solder paste on a substrate with a stencil.
   JIS Z 3284 specifies the substrate as Cu-clad laminate plate and J-STD-005 specifies it as frosted glass microscope slide. JIS Z3284

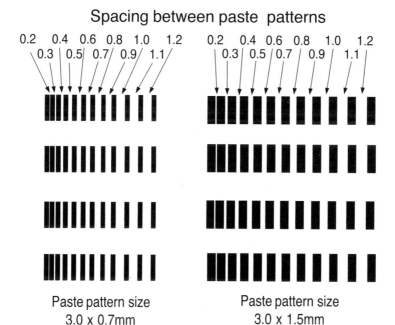

Figure 8.23   Layout of stencil opening for slump test.

specifies a stencil as a stainless steel plate of 0.2-mm thickness having two series of opening holes of 3.0 × 0.7 and 3.0 × 1.5 mm, whose spacing is arranged from 0.2 to 1.2 mm (0.1-mm step). The opening is shown in Fig. 8.23. J-STD-005 specifies it as a plate of 0.1-mm thickness having two series of opening holes of 0.63 × 2.03 and 0.33 × 2.03 mm whose spacing is arranged from 0.33 to 0.79 mm and from 0.08 to 0.45 mm, respectively. J-STD also specifies another type of stencil of 0.1-mm thickness having two series of opening holes of 0.2 × 2.03 and 0.33 × 2.03 mm whose spacing is arranged from 0.075 to 0.30 and from 0.08 to 0.45 mm, respectively.

    a.   Confirm that printing pattern is uniform in thickness without any solder ball separated from the printed pattern.

2.   Specimens should be stored under the following conditions:

    JIS-Z 3284

        a.   Room temperature for 1 hr.
        b.   150°C for 1 min (for Sn–Pb eutectic solder).

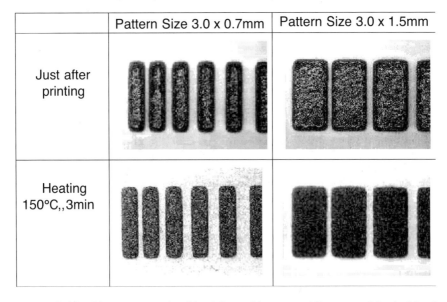

	Pattern Size 3.0 x 0.7mm	Pattern Size 3.0 x 1.5mm
Just after printing		
Heating 150°C,, 3min		

Figure 8.24 Slump test result of lead-free solder paste. (Courtesy of Senju Metal Industry Co., Ltd.)

J-STD-005

    a. 25 ± 5°C, relative humidity 50 ± 10%, for 15 ± 5 min

    b. 150 ± 10°C from 10 to 15 min

3. Check the minimum spacing that no solder paste is integrated.

The baking temperature for lead-free solder is not specified yet in JIS Z 3284 and J-STD-005. Although slump test under current test conditions may be irrelevant for lead-free soldering, the properties of slump of lead-free solder paste can be evaluated to some extent with current test conditions. Figure 8.24 shows a slump test result with a lead-free solder paste specified by JIS Z 3284. Solder paste doesn't show any slump after heating at 150°C for 3 min.

REFERENCES

1. S Fujiuchi. Collective screen printing for carrier bump and SMT pads. Proceedings of Japan IEMT Symposium, Omiya, Japan, 1995, pp 109–112.
2. S Fujiuchi. Evaluation of the thermal fatigue of QFP soldered joints with Sn–Zn–Bi based lead-free solder. Proceedings of 7th Symposium of Microjoining and Assembly Technology in Electronics, JWS, Tokyo, 2001, pp 447–450.
3. Senju Metals Industry Co., Ltd. R&D Report. Feature of Lead Free Alloys and Their Process Characteristics, 2000, p 6.
4. LL Bircumshaw. The surface tension of liquid metals: Part V. The surface tension of the liquid tin alloy. Phil Mag 17:181–191, 1934.
5. Senju Metal Industry Co., Ltd. R&D Report Lead-Free Solder for Environmental Protection. TC-01E, May 1998.
6. K Hagio. Improvement of printability for Sn–8Zn–3Bi Pb free paste. Proceedings of MES2002, JIEP, Osaka, 2002, p 171.

9
Wave Soldering

Tetsuro Nishimura

Nihon Superior Co., Ltd., Osaka, Japan

9.1 INTRODUCTION

The basic principles of wave soldering:

- contact of a fluxed solderable printed wiring board and component terminations with a continually refreshed surface of molten solder,
- heat transfer,
- wetting and flow, and
- drainage of excess solder

are the same for lead-(Pb)-free solder as for tin–lead (Sn–Pb) solder. The differences arise largely from the fact that:

1. The difference between the melting point of lead-free solders that are suitable for wave soldering and the maximum temperature that the printed board assembly and the wave soldering machine itself can accommodate without damage is much smaller than it is for Sn–Pb solder.
2. The wetting and spread properties of lead-free alloys suitable for wave soldering are not quite as good as those of Sn–Pb solder.

Despite these differences, it has been confirmed on many hundreds of wave soldering machines that have successfully soldered tens of millions of a wide range of printed board assemblies that lead-free wave soldering is a viable process that can produce reliable products economically.

The key to this success is process optimization, which takes into proper account the differences between lead-free and Sn–Pb wave soldering.

9.2 DIFFERENCES FROM WAVE SOLDERING WITH SN–37PB SOLDER

The most significant factor in wave soldering with lead-free alloys is that the superheat (i.e., the difference between the melting point of the solder and the solder bath temperature) is smaller, which is usually the case in wave soldering with traditional Sn–37Pb (Table 9.1). This means that process controls have to be tight enough to keep all joints within that smaller process window until wetting and fillet formation have been completed for all joints and, most importantly, in wave soldering, until the excess solder has drained off the joints.

9.3 CHOOSING A SOLDER BATH TEMPERATURE

In choosing a solder bath temperature, there are two considerations:

1. The minimum temperature at which, with the flux being used, the solder will wet the joint surfaces and flow to form the required fillet.
2. The need to raise the joint to that temperature during the few seconds that it is in contact with the solder.

One of the factors that determine the rate of temperature rise toward the required minimum soldering temperature is the temperature of the solder. If the solder bath is set to the minimum soldering temperature, then the laws of heat transfer mean that the joint will approach that temperature asymptotically and soldering will probably not occur within a practical contact time. The greater the superheat above that temperature is, the faster the joint

Table 9.1 Melting Temperatures, Typical Wave Soldering Temperatures, and Superheat of Solders

Alloy	Melting point (°C)	Typical wave soldering temperature (°C)	Superheat (°C)
Sn–Pb eutectic	183	250	67
Sn–Ag–Cu eutectic	217	250	33
Sn–Cu eutectic	227	255	28

surfaces reach the soldering temperature. Some degree of superheat is, therefore, usually employed to achieve fast reliable soldering.

For Sn–Pb solder, good wetting and flow can occur on copper (Cu) surfaces at temperatures down to 210°C so that at temperatures that the machine and assembly can safely tolerate (250–260°C), there is sufficient superheat to ensure rapid heating of joint surfaces to that minimum required soldering temperature.

For the Sn–Cu eutectic, for example, which has a melting point of 227°C, the minimum temperature for good wetting is around 240°C. With a solder temperature of 250°C, the rate of heating to the optimum soldering temperature is slower than it is for Sn–Pb soldering at the same temperature. This is one reason why solder contact time may have to be longer for lead-free wave soldering than it is when wave soldering with Sn–Pb.

9.4 DRAINAGE

It is well recognized that one of the key considerations in wave soldering, apart from achieving adequate wetting of all joint surfaces, is the removal of solder in excess of that required to form a fillet of the desired profile. Solder in excess of that required for fillet formation joint tends to form bridges (shorts) or icicles (Fig. 9.1).

The drainage (or "peel back") of excess solder from the joints occurs in the fraction of a second during which the joint exits the solder wave. Anything that reduces the mobility and fluidity of the solder at that point will result in an increase in the incidence of bridges. As with Sn–Pb solder, this drainage of lead-free solders is affected by machine parameters such as wave profile, wave stability, conveyor angle, and conveyor speed.

The ease with which the solder drains off a joint is also affected by the quality of wetting of joint surfaces with a well-wetted surface generally draining more easily. Flux activity plays a role in ensuring good wetting of the joint surfaces. Maintaining a low surface tension on the molten solder in the exit area flux also facilitates drainage. If fluxes are to have sufficient activity in this critical exit area, they must have an activation system that can survive the higher temperatures and longer times that might be required when wave soldering with lead-free alloys, and this point is discussed further in Sec. 9.12.

The special additional challenge for lead-free soldering, however, is maintaining the temperature of the solder in the exit area far enough above the melting point where the solder is fluid with no solid starting to precipitate. Although most of the lead-free solders being considered for wave soldering are nominally eutectic alloys (i.e., with a single sharp melting point),

Figure 9.1 Bridges (shorts).

examination of the microstructure of soldered joints made with lead-free alloys indicates that there is some precipitation of proeutectic primary phases. It would be expected that this early precipitation of solids would interfere with solder drainage and increase the likelihood of bridges. Thus, it is necessary to maintain as great a margin as possible above the melting point until the joint has left the wave completely.

9.5 MAINTAINING PROCESS TEMPERATURE

Because the temperature of the joints has to be held within a fairly narrow range right to the point of exit, it is important to minimize air drafts that may cool the solder at critical points (Fig. 9.2).

It is important that the heating system ensure that the solder in the wave is at the specified process temperature. Because the actual temperature of the solder in the wave may differ from the temperature indicated in the control system, it is worth measuring the temperature in the wave itself (Fig. 9.3).

In wave soldering machines in which the wave is switched on only when a board is about to pass over it, the bath heating system must be capable of holding the solder in the wave at the specified temperature when the pump is off. If there is any problem in that regard, it may be necessary to keep the

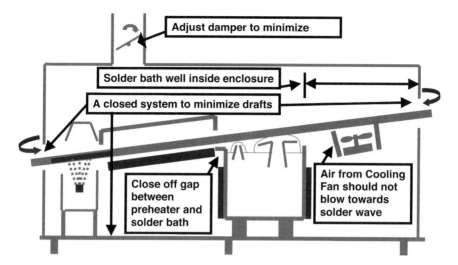

Figure 9.2 Minimizing the drafts that may affect chill process temperatures.

solder pump running slowly to keep the solder in the wave system at such temperature.

It is also important that the heating system be able to keep all parts of the solder bath well above the melting point of the solder. In some designs of heating systems, the greater heat loss from the corner can result in the solder temperature falling below its melting point with consequent partial solidification. Uniform solder bath temperature can be achieved with any well-designed system of heating elements, but this is probably easiest to achieve with external plate heaters.

9.6   PREHEAT

To compensate for the lower superheat, it is usually necessary to preheat the boards to a higher temperature than would be required when soldering with Sn–Pb. Typical preheat temperatures are set out in Table 9.2.

Because of the narrower margin between the melting point of the lead-free solder and the process temperature, it is even more important for lead-free solder than it is for Sn–Pb solder that the required preheat temperature is reached for all joints. Because of the different absorption characteristics of the various components that are used in printed board assemblies, it can be useful for the preheater to have more than one type of heat source (e.g., short-wave and long-wave length infrared and hot air).

Figure 9.3 Checking the temperature of the solder in the wave.

Table 9.2 Topside Preheat Temperature
for Wave Soldering (Sn–Cu or Sn–Ag–Cu
eutectic solder)

	Topside preheat temperature (°C)
Alloy	Sn–Cu eutectic
Single-sided	100–110
Double-sided	110–120
Multilayer	120–130

9.7   WAVE SOLDERING PARAMETERS

Because of the narrow process window, it is necessary to maintain process
parameters within a narrow range. Process parameters to which attention
needs to be paid include the following.

9.7.1   Contact Time

One of the most important process parameters is contact time and, because of
the slower wetting time and lower spread factor of most lead-free solders, this
usually has to be a little longer than would normally be used when wave
soldering with Sn–Pb.

Contact time is determined by the combination of contact length and
conveyor speed:

$$\text{Contact time (s)} = [60 \times \text{contact length (mm)}]$$
$$/[1000 \times \text{conveyor speed (m/min)}] \tag{9.1}$$

with contact length being determined by a combination of the wave profile
and the immersion depth.

Because of the narrow process window, it is important that contact time
be monitored. Contact time (and other wave soldering parameters) can be
monitored with dummy boards fitted with patterns of thermocouples whose
temperatures are recorded and analyzed by a computer. Such systems are
available from several vendors. Contact time can also be estimated by measuring contact length using a heat-resistant class plate marked with a grid.
This is carried on the conveyor with its bottom side in the same plane as the
solder side of the printed board assembly so that the dimensions of the wetted
area can be determined (Fig. 9.4).

To ensure that all joints experience the same contact time, it is important
that the contact length is uniform across the width of the board. This is
achieved by ensuring that the surface of the solder wave is flat and that the

Figure 9.4 Checking wave profiles and contact lengths with graduated glass plate.

plane of the conveyor is parallel to the surface of the solder wave. The height and the profile of the solder wave are determined by the evenness of the solder flow into the wave.

9.7.2 Conveyor Speed

Closed-loop control of conveyor speed is desirable to ensure that the speed remains constant, irrespective of the load of printed board assemblies on the conveyor.

9.7.3 Immersion Depth

Because of the generally lower spread factor of most lead-free solders, the filling of plated-though holes can be assisted by increasing the immersion depth to the maximum that can be tolerated without risking overflow of solder to the topside of the board.

9.7.4 Wave Profile

Because the basic dynamics of wave soldering is the same, a wave profile that has yielded good results with Sn–Pb should be suitable for lead-free solder.

The basic requirement is a steady flow of solder into the wave and a profile that will give the required contact time at the conveyor speed set. The conventional wisdom for some years has been to ensure that excess solder is not left on the joints; there should be smooth, stable laminar flow in the exit area. This is achieved with a pump-and-nozzle system that is designed to ensure stable flow with minimum turbulence and kept free of dross and other buildup that might interfere with the solder flow.

As with Sn–Pb solder, a first turbulent wave (chip) wave is recommended to assist the penetration of the solder into the termination area of chip components. However, because this turbulent wave also provides more effective heat transfer than a laminar wave, it is recommended that it be used even if there are no surface mount components on the solder side. For plated-through hole boards, the extra pressure of the turbulent wave can help penetrate lead-free solders through holes and can provide a better assurance that an acceptable topside fillet will be formed. In fact, there are turbulent waves of proprietary design (e.g., the Wertman wave) that can be used alone (i.e., without the second laminar wave that is traditionally required to eliminate bridges).

9.7.5 Thermal Profile

Because of the need to keep the printed board assembly in what is a fairly narrow temperature range with minimum total heat input, it is important that temperature drops between stages are kept to a minimum. Thus, the distance

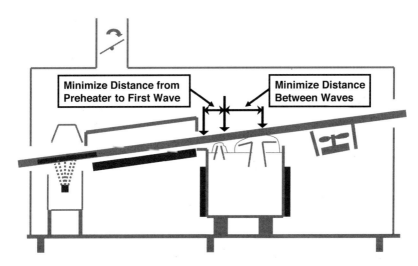

Figure 9.5 Maintaining temperature during the soldering process.

between the end of the preheater and the first wave should be kept to the minimum possible within engineering constraints and there should be no gap that would provide a channel for a cooling draft of air. Similarly, the distance between the first and second solder waves should be kept to a minimum (Fig. 9.5).

9.8 SUCCESSFUL SOLDERING OF PLATED-THROUGH HOLES

Although there are claims that better filling of plated-through holes is achievable with lead-free solders that contain Ag, the major factors appear to be process and design parameters. A key factor is getting the solder into the hole and, as mentioned elsewhere, using the maximum safe immersion depth; the use of a turbulent wave is useful in that regard.

Unless the hole wall and topside pad are at soldering temperature, wetting and flow cannot occur and a longer contact time achieved, for example, by slowing the conveyor allows more time for that to be achieved (Fig. 9.6).

Even providing a larger area for heat transfer to the lead in the through hole can make a difference. Figure 9.7 shows how leaving a longer lead length can have an effect on topside fillet formation.

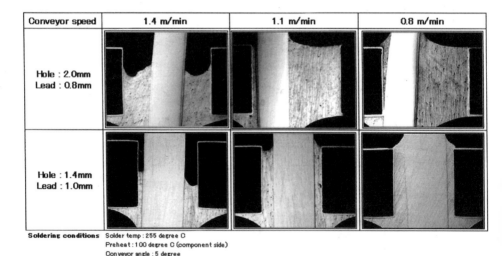

Conveyor speed	1.4 m/min	1.1 m/min	0.8 m/min
Hole : 2.0mm Lead : 0.8mm			
Hole : 1.4mm Lead : 1.0mm			

Soldering conditions Solder temp : 255 degree C
Preheat : 100 degree C (component side)
Conveyor angle : 5 degree
Flux : Rosin Mild Activated (RMA) Type

PC Board FR-4 1.6mm t
OSP finish

Figure 9.6 Effect of conveyor speed on hole filling.

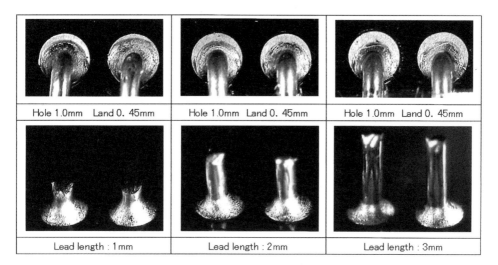

Hole 1.0mm Land 0. 45mm	Hole 1.0mm Land 0. 45mm	Hole 1.0mm Land 0. 45mm
Lead length : 1 mm	Lead length : 2 mm	Lead length : 3mm

Figure 9.7 Effect of lead length on topside fillet.

9.9 CU LEACHING

Sn, the major ingredient in all of the lead-free solders suitable for wave soldering, dissolves Cu rapidly at soldering temperatures. This aggressiveness is thought to be related to the fact that Sn readily forms intermetallics with Cu. Cu leaching is recognized as a problem even with Sn–Pb solder even though the tendency is considerably reduced by the fact that the Sn is diluted with lead, which has little tendency to dissolve Cu. The high Sn content of the lead-free solders means that they are potentially more aggressive toward Cu than the Sn–Pb solder, although that effect may be moderated by the level of Cu in the alloy and the presence of other alloying additions. Ag, which also readily forms intermetallic compounds with Cu, appears to increase the tendency of the solder to dissolve Cu and this has implications for the management of a wave soldering bath. There is also some evidence that the aggressiveness of the Ag-containing lead-free solders toward Cu may have implications for the reliability of printed board assemblies soldered with them (Fig. 9.8).

9.10 MANAGING THE SOLDER BATH COMPOSITION

Although the issue of Cu contamination is well known in wave soldering with Sn–Pb solder, it has to be viewed from a different perspective in lead-free

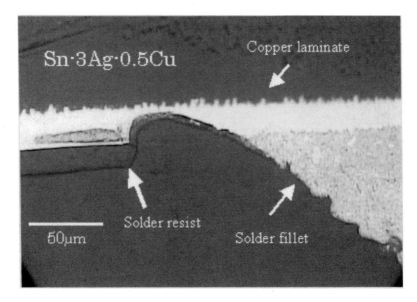

Figure 9.8 Copper erosion of tracks, pads, and hole wall by Sn–3.0Ag–0.5Cu solder.

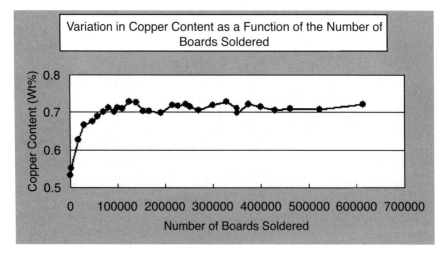

Figure 9.9 Managing the copper content by use of a top-up alloy with a lower copper content.

soldering because Cu is one of the key alloying additions. In the Sn–Cu eutectic, the Cu content must be kept within the range 0.6–0.85% if consistency of results is to be maintained. In the Sn–Ag–Cu eutectic, the Cu level must be kept within 0.5–0.9% for consistent performance. For some Sn–Ag–Cu alloys, there is the additional legal issue that the alloy, if the Cu content rises too far from the printed board assemblies soldered with it, may be in breach of a patent that covers the composition of the solder in the joints as well as the solder added to the solder pot.

As in Sn–Pb solder, the rate of Cu dissolution from the board is affected by the level of Cu already in the solder so that an alloy with at least the eutectic level of Cu appears to dissolve further Cu significantly more slowly than alloys with a Cu content significantly below the eutectic. As noted earlier Ag, which by itself tends to dissolve Cu, seems to increase the tendency of a solder to dissolve Cu from boards and components, and this can make it difficult to keep the Cu content at an acceptable level.

Provided the rate of Cu dissolution is not too great, it has been found possible to maintain the Cu at the specified level by the use of a top-up alloy that contains less Cu (Fig. 9.9).

9.11 SOLDER BATH EROSION

Sn is aggressive toward ferrous alloys as well as Cu and because the lead-free alloys suitable for wave soldering are all high in Sn, there is a greater tendency toward erosion of the stainless steel that has commonly been used for the fabrication of solder pots, solder pumps, and wave formers. Indeed, there is already considerable experience of solder bath erosion (Fig. 9.10).

The aggressiveness of high-Sn solder toward stainless steel seems to be increased if Ag is also present. The basic Sn–Cu eutectic is less aggressive than the Sn–Ag–Cu eutectic, but at sufficiently high temperature (e.g., 280°C and above), some erosion will occur. Such high temperatures have been used only in an attempt to get satisfactory results with the basic Sn–Cu eutectic. It has been found that if a trace of nickel is added to Sn–Cu [1], it can be used as practical wave soldering alloy at temperatures no higher than 260°C, at which erosion of stainless steel seems to occur much less than Sn–Ag–Cu.

Cases of erosion have often been found to be associated with local overheating. This can sometimes be a problem with stainless steel sheathed immersion heating elements where a "hotspot" with a surface temperature around 300°C can sometimes occur. At such a temperature, erosion can proceed quickly. Local overheating that could provoke erosion is less likely to occur with properly designed and externally fitted plate-heating elements.

Figure 9.10 Solder bath nozzle erosion by Sn–Ag–Cu alloy.

The resistance of stainless steel to erosion by high-Sn alloys is increased as the nickel and chromium contents increase and such higher-grade materials should be used. A higher level of resistance can be provided by proprietary coatings, but consideration has to be given to their vulnerability to mechanical and thermal damages.

9.12 FLUXES

The role of the flux in lead-free soldering is the same as that in Sn–Pb soldering, that is:

- Removing oxides from the board and terminate surfaces
- Removing oxides from the solder surface
- Providing a barrier against further oxidation of these surfaces
- Facilitating heat transfer into the joint surfaces
- Acting as medium for metal ion transfer as part of the wetting process.

These functions are important in two parts of the wave soldering process. These processes are important in the first stage of achieving wetting of all the joint surfaces. They are important also in the final stage in the exit area of the wave where solder in excess of that required to form the required fillet profile drains back into the wave.

To do both of these jobs, the flux must be able to survive in a mobile, active condition until the exit area. The generally higher temperature and longer contact times required with lead-free solders impose greater demands on the flux than is usually the case in Sn–Pb soldering.

Higher solid content resin-based fluxes with a solid content of more than 10%, such as are widely used by the Japanese consumer electronics industry, seem well able to meet the more severe demands of lead-free soldering. Many companies using such fluxes have found that they can use the same flux with lead-free solder as they have been using with Sn–Pb solder.

Many low solids "no clean, no residue" fluxes used in Europe, North America, and Asia outside Japan for Sn–Pb soldering may not be suitable for lead-free soldering. With solids content typically less than 5% and sometimes less than 2%, the activation systems in these fluxes may not survive to the thermal profile. The result would be seen as bridges and icicles in incomplete through-hole filling and topside fillet formation. The use of a complete or partial nitrogen atmosphere can reduce the stress on these fluxes so that satisfactory results can be achieved. If soldering in air is to be continued, these low-solid fluxes have to be reformulated with more robust activation systems without compromising properties that affect reliability such as surface insulation resistance and electromigration. Such fluxes have already been developed and will provide the basis for successful lead-free wave soldering in air.

REFERENCE

1. T Nishimura. Behaviour of Wave-Soldered Joints by Sn–Cu System Lead-Free Solders. Trans Inst Electron Inf Commun Eng Sect C J85-C(11):961–967, 2002.

10

Soldering Inspection and Design for Lead-Free Solder: Soldering Innovation by Trinity of Design, Process, and Inspection

Masao Hirano*

Omron Corporation, Tokyo, Japan

10.1 GENERAL VIEW OF SOLDERING INSPECTION

The advancement of soldering technology has been remarkable in the last two decades. High-density packaging has been in the forefront of electronics packaging, especially for notebook-type personal computers and mobile phones. Lead-free soldering is now taking the position of the conventional leaded solders and much research effort has been exerted since around 1998 in commercial fields. In addition, in order to transform soldering processes to lead-free soldering, miniaturization of the soldering equipment and changes in the production infrastructures were started in 2001. New soldering technology has evolved in this way with surprising speed.

Current affiliation: Kyoto Profeature Adviser (Technology Issues), Kyoto, Japan.

The soldering defect, however, still exists in actual production. Lead-free soldering sometimes becomes difficult because of the high melting point of lead-free solders. Defect-free production must be achieved before lead-free soldering becomes more common in production.

How does soldering become difficult by lead-free soldering? The difficulty appears as follows:

1. The proper temperature range of reflow narrows because the melting point of the solder rises without improving the heatproofing of components.
2. Poor wetting of solder occurs, which is the nature of lead-free solders. Even though a solder fillet becomes small due to poor wetting, joining strength for lead-free solder is much better than that of leaded solders.
3. During inspection, the inspection standard of the quality item changes because the fillet shape of joint varies.

In the transition of leaded soldering to lead-free soldering, one should understand the generation mechanism for each defect.

Because roughing of the surface of a solder fillet occurs for most lead-free solders, inspection sensitivity may become worse than for a leaded solder fillet. In practical use, a lead-free solder sometimes causes larger scatters in fillet shape than that of a tin–lead (Sn–Pb) solder. This variation seems to cause the degradation of the joint reliability or the production yield. Therefore, one needs to review the land design of a printed wiring board (PWB), print mask design, manufacturing, and soldering inspection. The fillet design, which can improve inspection sensitivity, should be performed to secure reliability and manufacturability for industrial use.

Therefore, the "ideal way of inspection in lead-free soldering age" is interpreted as the "ideal way of PWB design and process design." First of all, the integration of the design and the process to provide a good quality of products is described. Actually, when the author started mass production using lead-free solders, an inspection machine was effectively utilized in confirmations of new process conditions and new PWB designs.

10.1.1 Purpose of Soldering Inspection

In the inspection of a soldered PWB, the soldering inspection is usually carried out because of accuracy, generality, and ease of the inspection. The most typical inspection is that of a solder fillet after the SMT process. As for an electrical circuit test, its circuit function is often confirmed with a specific tester after assembly.

Figure 10.1 shows the sequences of inspection with the purpose and the system. The purpose of the process inspection is: "Quality can be given by

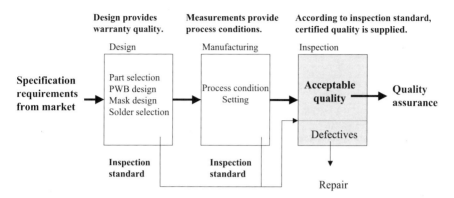

Figure 10.1 Purpose and system of inspection.

the design standard and the process standard, and only the quality item of the performance characteristic more than the inspection standard is supplied to the next process." The desired quality level comes from the specification requirements from the market. The output of this system is, of course, a guarantee of a designed quality of production. Figure 10.1 shows that PWB designs provide warranty quality, and that process condition control provides a crowded quality. Inspection standards from design and process supply only certified quality up to the next process. Therefore, the acceptable quality and defects can be distinguished by referring to a given inspection standard.

In general, the quality of inspection is defined both by the probability of defective goods being mixed into the quality goods (called rate of underestimate) and by the probability of the quality goods being mixed into defective goods (called rate of overestimate). Because defective soldering is easily repaired, the definition of the latter is not so problematic. As for the former, it is likely to overestimate to put the quality of conformance on the safe side.

10.1.2 Variety of "Measurement and Inspection" Used in Soldering

The factors that guarantee the quality of soldering include the following:

1. Design elements, which include components selection, board design, screen mask design, solder selection, etc.
2. Manufacturing elements, which include conditions of machine group.

Figure 10.2 shows the system of "measurement and inspection" used in the practical production of soldering. Taking a general view of the flow

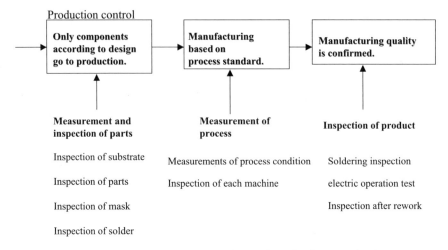

Figure 10.2 "Measurement and inspection" for jisso production.

chart, in the first step, the measurements of the size and characteristics lead to the confirmation of a screen mask, PWB, components, and solder turning to correct materials, just as designed. Next, measurements of the process conditions such as printing, mounting, and reflow/wave, and inspections of the qualities of manufacture at a middle stage of SMT processes are done as they proceed right on process standards. After such an SMT process, solder fillet inspection and the circuit response test are carried out to supply only the quality item to the postprocessing. As for a defective PWB, when the inspector finds it, the inspection is again carried out after repair.

The most typical inspection is that of a solder fillet after SMT process with a variety of measurements and inspections. This stage of inspection is carried out not to transfer a defective PWB to the next assembly process. Off-line inspection is not common, but in-line inspection is valuable in soldering. According to recent miniaturization, the utilization of a device measuring screen-printed fine paste pattern in three-dimensional (3D) imaging becomes very popular, as shown in Fig. 10.3. This is the first step toward process improvement.

10.1.3 Types of Soldering Inspections

In most cases of the current soldering process, inspection after the SMT processes is very much popular. Typical inspection methods are listed in Table 10.1. The method of PWB inspection after soldering can be divided roughly from the inspection purpose into judgment of "quality of solder

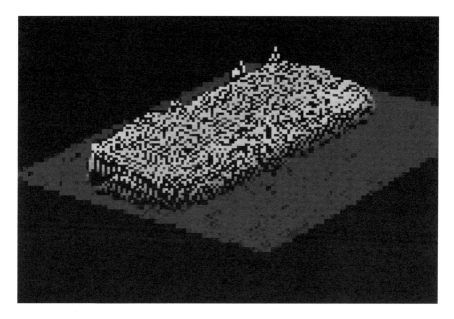

Figure 10.3 The 3D measurement of printed solder shape using light section method.

Table 10.1 Method of Typical Soldering Inspection

Inspection purpose	Object of inspection	Inspection method
Quality of soldered joint	Fillet shape	Visual inspection
		Optical displacement: 2D section, 3D shape
		Scanning type: 2D section, 3D shape
		Image processing: 2D plane
		Color highlight: 3D shape
		X-ray penetration: 2D plane
		CT: 3D shape
	Electrical connection	Electrical inspection
	Others	Heat flow rate
Quality of circuit operation	Circuit operation	ICT
		BST

joint" and "quality of circuit function." Thus, in the quality judgment of the soldered joint, there are "methods of judging fillet shape on PWB" and "methods of judging the electrical connection of PWB."

In the method of judging fillet shape, the image processing method is one of the common methods used in inspecting the fillet shape. The reason in using the image processing is the general feature corresponding to various components. Because of the miniaturization of components and because of the expanding variations of components, the generality of an inspection tool becomes more and more important. Because soldering inspection is basically the inspection of all joining points, inspection sensitivity to various objects (e.g., fine components to complex ones) and inspection speed become two of the key performances.

As for "shape judgment," the quality of joining is judged from a fillet shape on the assumption that the solder and the electrode are wet well and that the metallic interconnection is accomplished at each joining point. Securing the shape of a solder fillet actually secures not only the electrical current path but also the thermal cycle stability of a joint. The method of recognition of solder fillet shape includes scanning by a spot or a line of light and processing its image. Image processing can produce the 3D information of a solder fillet through color change by catching the reflection from a solder fillet irradiated by three color light sources of R, G, and B. This method becomes very popular as a "color highlight method." In this imaging, although observation itself is a two-dimensional (2D) measurement, multiplied images by inclination of a solder fillet to the color light enable the 3D inspection.

Area array components with bump joints of CSPs and BGAs cannot be inspected by optical imaging methods. This is because the bumps under the package cannot be recognized by light. For inspecting components with bumps, the x-ray inspection has begun to be used together with optical imaging.

In the next step, an electrical connection measurement is described. Although circuit test is important, there are still some problems in inspection. Basically, the inspection must be the simple GOOD/NG judgment, and the current or electrical resistance is measured. Because the resistance of solders is relatively small compared with the sum total of the circuit resistance of joint structure, it becomes difficult to distinguish a faint change of electrical signal originating from only one soldered point against a whole circuit. Thus, this inspection cannot be a simple GOOD/NG judgment. It sometimes happens that mechanical contact behaves like a perfect soldered junction. Thus, electrical circuit measurements generally have serious problems with poor inspection sensitivity in distinguishing the quality item and the defectives.

In-circuit test (ICT) is often used as a high-performance PWB. Again high-density packaging makes it difficult because there is not enough room for installing the test pin on a PWB. The high cost of test pins is another problem. One of the recent trends is the effective function test method of "boundary scanning test (BST)," in which a small test circuit is built into an IC. By BST, test pin numbers can be effectively reduced.

The importance of the electrical circuit tests is that it guarantees the circuit operation of the module production, as the module design becomes a common method.

10.1.4 Evolution of Inspection Methods

Along with developments of lead-free soldering, inspection changes from the inspection of an entire SMT process to individual process inspection. Figure 10.4 shows the evolution of the inspection method for SMT of lead-free soldering. Because of the technological difficulties of lead-free soldering, a screen printer, a mounter (chip shooter), and a reflow furnace in an SMT sequence each undergoes inspections individually. Screen printing shows the tendency of increase of printing defects due to miniaturization of components. The mounting shows the tendency of increase of shooting miss due to miniaturization and increases in miss multisetup. As for reflow, defects increase by poor wetting of lead-free solder and miniaturization of components. Thus, each step in the production has its own defect formation; thus, putting an inspection tool just behind each machine is one of the best ways.

The adoption of inspection after screen printing has become especially common. A fine solder paste printing pitch of 200 µm or less in width has been needed for the bumping of CSPs, BGAs, and 0603 chip component. The printing pitch of 200 µm or less only involves up to four to five particles of solder particles in one line. The transcript rate decreases by 20% or more even

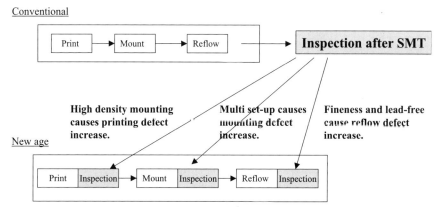

Figure 10.4 Evolution of inspection method for SMT upgrade.

in the case of one line miss. Moreover, it is a background reason why the inspection spreads in screen printing and why careful screen printing helps produce high yields for CSPs. Please remember that CSPs cannot be inspected by image processing after mounting.

10.2 NEW INSPECTION METHOD FOR LEAD-FREE AND HIGH-DENSITY PACKAGING

10.2.1 Requirement for Lead-Free Solder in Inspection

Soldering inspection must be modified to lead-free soldering from the conventional one suitable for Sn–Pb eutectic solder. The modification is based on the difference between lead-free solders and Sn–Pb solders. The differences are as follows:

1. Difference in physical property
2. Difference in quality or shape of a solder fillet
3. Difference in optical property of solder and flux surfaces.

Where does the change occur for lead-free solders? Table 10.2 shows the characteristics of lead-free solders and their influence on soldering feature.

Table 10.2 The Property of Lead-Free Solders and the Influences

Property of lead-free solder	Influence	Countermeasure of inspection machine
Decrease in wet spreading, slow wetting	Change in fillet shape Variation increase of shape Partial wetting on electrode (dewetting)	Change in inspection standard
Change in reflow heating condition	Temperature rise of parts (wicking at leads of QFP)	
Change in fillet surface luster	Sensitivity decrease of inspection	
Replacement of part electrode	Variation increase of shape Wetting change to electrode (dewetting)	New algorithm
Combination of part electrode and solder	Fear of reliability decrease	

Necessary correspondence for the inspection machine is also listed in the table. The largest influence appears in reflow condition such as temperature rise. Change and severe management of measurement of reflow temperature are required. For an inspection criterion or machine, those changes appear as electrode material change, or a change in fillet shape and its glossiness. Especially, the change of reflection of a fillet surface influences inspection sensitivity.

It is possible to correspond to the shape change and the difference of the fillet of joint by a "review of inspection standard" and "fillet design that facilitates inspecting" described later. Moreover, new inspection logic can be developed for the improvement of inspection sensitivity.

What is the influence of the rough surface of a solder fillet on image processing? Lead-free soldering makes changes: (a) in reflections on solder fillet surface, and (b) in a fillet shape. Especially, one needs to be careful that the decrease of reflection on a fillet surface seriously influences inspection.

Principles of Various Visual Inspection Equipment and Reflection Characteristics on Solder Surface The visual inspection machine available in the market is classified into six different illumination methods:

1. Level illuminations using three color light sources of RGB
2. Level illuminations using white light sources
3. Level illuminations using LED (monochromatic)
4. Coaxial head illumination using LED (monochromatic)
5. Line scanning illumination using laser light
6. Projection geometrical pattern of interference-induced stripe.

Moreover, the optical receiver that catches the reflected light is classified into three:

1. Area imaging receiver using area CCD
2. Line scanning receiver using line CCD
3. Arranged according to receiving optical angle receivers.

Although there are various methods with different illuminations and optical sensors, most methods have in common the following: (a) in the sensing system, light is irradiated on a solder surface and a component; (b) reflected light from a solder surface is caught with a sensor; and (c) the inclination of a surface is recognized as strength of reflected light. Therefore, when the luster of a solder surface changes, misinterpretation cannot be avoided in every method.

Principle of Color Highlight Inspection The most typical inspection method is the color highlight method. The influence of a rougher surface of lead-free solders becomes a key factor.

The principle of the color highlight method is shown in Fig. 10.5. Three color light sources of the annulus ring of R, G, and B changing angles into the solder surface on a PWB are irradiated. A color image of the diffused reflection from the curved solder surface is obtained with a CCD camera right on top for inspection. The solder surface profile of three dimensions can be converted to RGB patterns of two dimensions. The shape information in three dimensions is converted into the hue information of two dimensions. The function of the conversion is shown in Fig. 10.6. In this method, inspection sensitivity becomes high because the angle of the solder surface correlates to irradiated hue, and the influence of the change in reflection light strength is small.

As shown in Fig. 10.6, because the RGB light sources are arranged in an order far from the PWB, the reflection lights of R comes from a top flat slope, those of G come from the middle easy slope, and those of B come from the bottom steep slope. Thus, one can obtain pattern information corresponding to reflection positions and solder surface conditions.

Influence of Luster Decrease and the Correspondence in Color Highlight Inspection   Lead-free solder is likely to have a rough surface because of its solidification feature, as mentioned in Chapter 2. The influence of the decrease in luster of a surface of solder on the inspection should be understood properly.

Figure 10.7 shows one of the typical comparisons between Sn–Pb eutectic solder and lead-free solder, Sn–Ag–Cu, and its inspection images for a chip component. Fig. 10.8 shows the same set of photographs for a QFP.

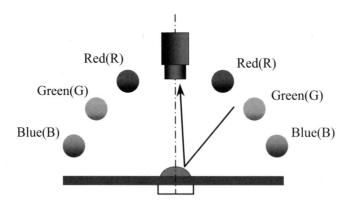

Figure 10.5   Inspection principle of color highlight method. Irradiating color lights of R, G, and B that set with the specified angles to a PWB. Reflected light from the solder surface is detected by a CCD camera set perpendicular to the PWB.

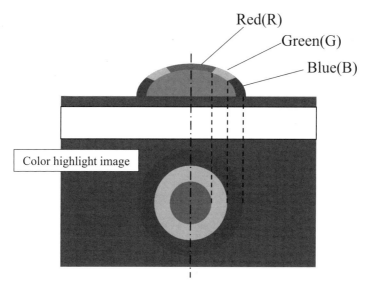

Figure 10.6 Color highlight image that inspects curved surface. It is classified into a color highlight image consisting of blue, green, and red from substrate sides to the perpendicular line according to the curved surface of solder.

The surface of Sn–Pb eutectic solder is smooth and has luster. The reflection light from this surface is strong like the specular. The surface of the lead-free solder fillet becomes rather rough, like a satin-finished surface. The reflection from the lead-free solder surface becomes diffused and the light strength decreases. When the degree of roughness of a surface is especially strong, the color components of RGB in the reflection light become near equal. Then, the color pattern tends to be white because diffuse reflection becomes remarkable. However, the relation of the angle to the hue of the solder surface is kept even if the color becomes whitish. When the diffuse reflection becomes significant and when the color patterns whiten too much, the inspection sensitivity may decrease. Therefore, to correspond to the effect of the change in surface roughness, the color emphasis processing to create clear color pattern images should be employed. The color emphasis processing is the processing that emphasizes color by removing the white component.

A concrete example of color emphasis processing is shown in Fig. 10.9. A square center area has a rough surface, a satin-finished surface. This image processing sequence is explained in Fig. 10.7. The right figure of Fig. 10.10 shows that the RGB color intensities after color emphasis. The whitish area at the central square in Fig. 10.9 almost disappears in the right image. Thus, the

Sn-Pb   Sn-Ag-Cu

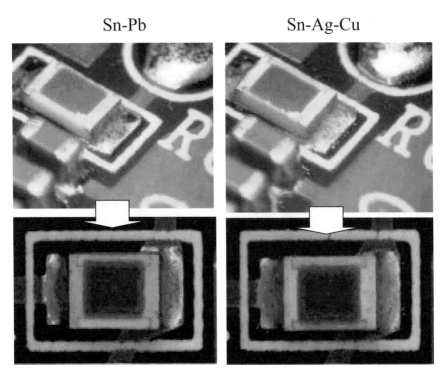

Figure 10.7 Influence of solder roughness on color highlight image in case of chip components.

recognition of RGB colors can be improved to a clear color pattern by a simple color emphasis processing [5].

Influence of Fillet Shape Change in Color Highlight Inspection and Correspondence The influence of fillet shape change and the countermeasures of the inspection are described.

The inspection corresponding to various fillet shapes should be also examined. The correlation between various fillet shapes, for which the amount of solder may change in using a eutectic solder and a lead-free solder, and the inspection sensitivity are evaluated. It can be confirmed that the color highlight images both for a eutectic solder and a lead-free solder have almost the same hue distribution, and there is no problem in inspection to the change of solder fillet shapes.

Because wetting of solder varies at the top of the leadframes of QFP, a new inspection logic that can distinguish the differences of wetting degree was

Sn-Pb   Sn-Ag-Cu

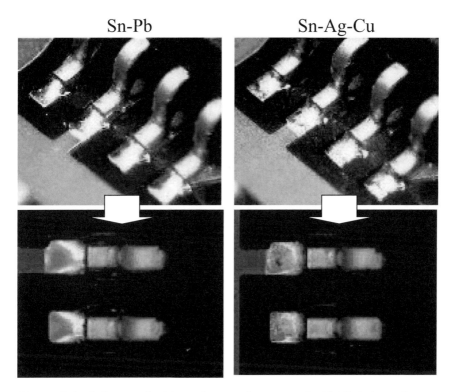

Figure 10.8 Influence of solder surface roughness on color highlight image in case of QFP.

developed. By this evaluation, defectives induced by "overestimation" and "underestimation" can be greatly reduced.

As for changes of the terminal plating of components, which may cause worse wetting, it is possible to avoid the influence by changing the inspection standard within the range where reliability can be allowed. For instance, a lower fillet height is enough for most lead-free solders to achieve the same level of joining strength or reliability as a Sn–Pb solder.

10.2.2 Fillet Design That Facilitates Inspections

A fillet shape design greatly influences the joint reliability in soldering production and one needs to optimize this for lead-free solders. It may be necessary to review a PWB design. The PWB designing involves both land design and mask design. By using the conventional color inspection machine, it becomes possible to apply for the color highlight method lead-free solder

Without color emphasis processing
Conventional processing

With color emphasis processing

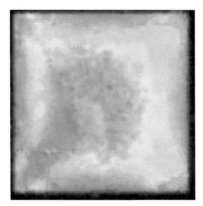
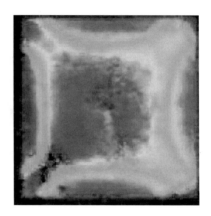

It looks white.   Red appears.

Figure 10.9   New color highlight image solving problem with surface roughness of solder.

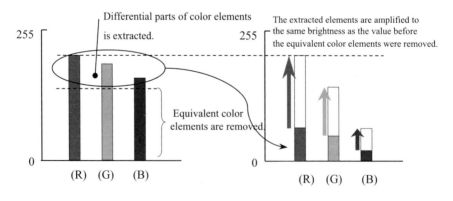

Figure 10.10   Countermeasure to prevent the influence of solder roughness on color highlight image.

assembling without changing the image processing. The techniques can be summarized in three points:

1. Adoption of fillet design that facilitates inspecting
2. Definition change for lead-free solder reflection
3. Acceptance/rejection level change corresponding to increased variation of fillet shapes.

"Fillet design that facilitates inspecting" was especially effective compared with other two measures. Figure 10.11 shows the outline. To facilitate inspection, it is assumed that a thin fillet of valued-added design with enough industrial reliability should have certain necessary length in its front fillet as it can be easily inspected. For the example of a chip component shown in Fig. 10.11, the appropriate front fillet length is about 0.2–0.35 mm. This value is derived from the confirmation of reliability, manufacturability, and inspectability.

If the fillet has a plateau, an inspection image is not formed. The fillet shape with an excess flat part is generally undesirable from the viewpoint of thermal fatigue endurance for small components.

10.2.3 New Inspection System

Let us discuss individual inspection in unit processes of SMT. There is a movement to individual inspection in unit processes of screen printing, mounting, and reflow soldering according to the sequences in SMT because of the difficulty of only one inspection after all process passages in the development of fine components and fine pitch as well as lead-free soldering.

	NG	OK	NG
Viewpoint	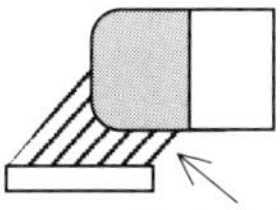 No-back fillet	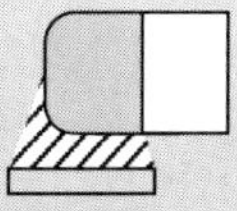 With back and front fillets	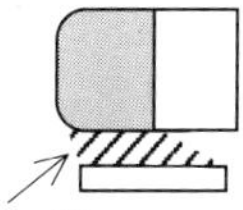 No-front fillet
Reliability		Appropriate	Weak to stress, thermal fatigue
Manufacturability	Floating, standing, or θ gap after reflow	Appropriate	
Inspectionability		Appropriate	Visual inspection is difficult.

Figure 10.11 An example of fillet design that facilitates inspection for industry use.

The stabilization requirement in production needs an inspection for each element in SMT.

Table 10.3 summarizes the details of each purpose of individual inspection in SMT process, the inspection characteristic, and the inspection method.

In the first stage, screen printing of solder paste and its configuration is evaluated in accordance to "maintenance of the transcript rate" and "prevention of the human failure of setup of printing." Printed transcript rate is also judged to keep an appropriate production speed, especially for fine components, and deterioration judgment in the transcription during continuous printing becomes important. A viscosity increase of solders must be avoided. A lot of defects ("graze, lack, sober, and bridge, etc.") must be ob-

Table 10.3 Inspections in Unit Processes for SMT

Inspection process	Inspection purpose	Inspection characteristics	Inspection methods
Screen printing	Keep high print transcript rate Judgment of setup suitability	Printed patterns Shape of solder paste	Visual inspection Optical displacement: 2D section, 3D shape Scanning type: 2D section, 3D shape Image processing: 2D plane Color highlight: 3D shape
Mounting	Positional accuracy maintenance	Presence of correct components	Image inspection position information
	Discovery of arrangement error at setup	Arrangements	Character and shape
Reflow	Judgment of suitability of soldering (correct fillet shape)	Fillet shape Arrangements	Visual inspection Optical displacement: 2D section, 3D shape Scanning type: 2D section, 3D shape Image processing: 2D plane Color highlight: 3D shape

tained as information in a suitable 2D $(X-Y)$ image. However, the information in the height direction (Z) is desired to add to the performance in judging the screen printability evaluation as shown in Fig. 10.2. At this moment, only the 3D measurement of off-line is possible for use to check screen printing conditions primarily to inspect speed for 3D measurement. Although many 3D inspections, such as a linear inspection using a light spot and an optical cutting inspection using a laser beam, are available in the market, most of them are in the category of 2D inspection.

In a mounting process, the position and the direction of components are evaluated for the purposes of "discovery of angular arrangement error of components" and "accuracy of position where components are installed." The accuracy management in mounting increases its importance along with the movement of fine pitch mounting. By the observation of the change of the gap among components, it is expected to inform the maintenance opportunity of a mounting machine at a suitable period.

Because a reflow process is the last process of SMT, all defects including poor wetting, floating, standing, θ gap, voids/cracks, and solder balls must be found.

When the inspection machines are arranged in all of these three processes, one can expect to achieve the following:

1. Manufacturing becomes steady.
2. Analysis of defect origins becomes easy.
3. Each effect is comprehensible corresponding to tweaking factor in a process.

Production using this inspection will not cause the lot out because human error and machine change at the setup can be stopped from the first piece of production in the inspections according to the processes.

10.3 SOLDERING INNOVATION OF THE 21ST CENTURY USING MEASUREMENT AND INSPECTION

Along with the new wave of soldering in the 21st century brought about by lead-free soldering and fine pitch packaging, a production form is changing rapidly. The engineering of material, manner, and machine (i.e., 3M engineering) will be fully utilized to evolve new soldering technology and to stabilize it in a short period. We believe that new sensing technology brings about the developments of this soldering evolution.

The idealized flow chart of the total inspection system in the future is schematically shown in Fig. 10.12. A new in-line inspection system is supported with off-line measurements and with the technology of 3M engineering. Here, 3M engineering theory takes on a role in the preliminary

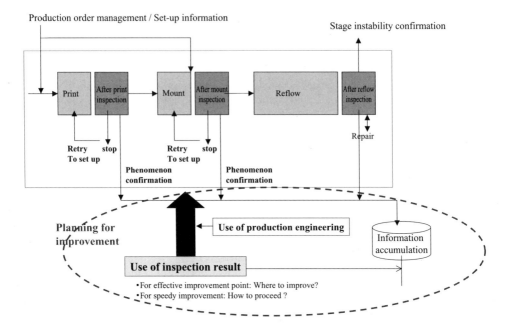

Figure 10.12   New individual inspection system for SMT.

evaluations of reliability, manufacturing, and new component installations. For new components and a new machine or process adoption, the intense initial evaluations of reliability, manufacturability, and inspectability are required. It can be said that the 21st century is the Renaissance Age of Soldering. We believe that the evolution can be accelerated by the trinity of material, manner, and machine; by the trinity of the design, process, and inspection; and by the instrumentation technology of a new soldering as well.

APPENDIX: LEAD-FREE SOLDERING PRACTICE IN OMRON

10.A1 DEFECT NOT TO BE VANISHED, AND THE CAUSE AND THE COUNTERMEASURES

10.A1.1   Process Function and Manufacturing Stability of Soldering

In soldering, there are three typical soldering methods of reflow, wave, and hand/robot. Reflow is a suitable technique for high-density packaging and is a representative manufacturing method. Nowadays, reflow accounts for about

60–80% of all soldering production. As for yield, it is possible to reach the rate of the quality item of 99.9999%, and the manufacturing stability is excellent after suitable optimization. On the other hand, wave soldering and hand soldering are the methods used to connect the insert components to through halls on a PWB. Wave soldering is suitable for PWBs equipped with large components such as capacitors of power supply circuits. The yield of those soldering methods is not so high, approximately up to 99.9% at the highest.

Through the observation of three different soldering processes, it becomes possible to understand the reason why the yield is greatly different among them. The soldering behavior of a fillet formation of each soldering method is compared in Table 10.A1. The characteristics of various processes such as solder supply, wetting, and fillet formation can be clarified from the behavior of the solder. The process function in reflow is very simple, which can be attributed to the high yield in actual production. Such soldering behavior can be measurable. The evaluation of manufacturing goodness that cannot be measured only by the yield became possible by this approach.

10.A1.2 Reasons Why Defects Cannot Vanish

The defect will not vanish at all in the actual production even though the stability of SMT manufacturing is excellent. In the age of lead-free soldering, we want "zero defect production" by all means.

Table 10.A1 Criteria of Soldering

soldering method / Process function	Reflow	Wave	Hand soldering
Kind of solder	Paste	Ingot	Wire
Solder supply method	Print	Soaking	Pour
Soldering Process observation ↓ **Behavior of solder at forming fillet**	Put the solder of necessary amount to necessary position Melt it there.	Dip melt solder and peel-back against surface tension of melt solder	Give heat from tip Pour solder to the narrow space of tip and the terminal
↓ **Definition of function characteristic**	■ Transcript ■ Coalescence ■ Interfacial tension ■ Surface tension	■ Immersional wetting ■ Surface tension ■ peel-back	■ Heat transmission ■ Flow sped of flux ■ Flow speed of solder

■ **Simple process function provides stability of SMT.**

The author did improve the process to make no defect before starting mass production using lead-free solders. The fact that "the material and the reflow are black boxes" is noticed in the process system of SMT. Solder paste is described by a chemical specification, Therefore, the paste is not well modified to necessary characteristics from the processes of printing, chip shooting, and reflow heating. The paste is exactly a black box for processes. The reflow is another black box. In fact, the reflow is done in the box of the furnace so that the process is not seen. Especially, the relation between the paste characteristics and the reflow heating conditions is not well understood (refer to Fig. 10.A1).

Moreover, there was difficulty of understanding easily which process made which defect, because current solder external inspection did the inspection of the entire SMT, which is the other black box. Although it is rather easy to see the phenomenon in print, it is difficult to catch phenomena of chip shooting in time for operation at high speed. The movement of the machine sped up more and become a black box as a recent trend.

The abovementioned seem to be the reason why the defect will not vanish.

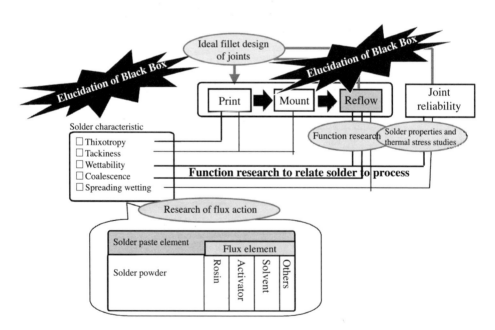

Figure 10.A1   Black box of SMT and researches to relate solder to process functions.

10.A1.3 Opening the Black Box in the Reflow

The outline of the research that opens the black box of the SMT is shown in Fig. 10.A2. "Research to which the material and the process function were related" proceeded according to criteria of soldering in Table 10.A2. The process was stabilized with the functional study for goodness of manufacturing and the investigations of the causes of defective generations. Especially, it gave priority to examining the relation of the reflow and the solder paste.

The complex behavior of soldering procedures that occurred one after another in reflow was illustrated in Fig. 10.A3. In Fig. 10.A4, the behavior through the heating process was shown on the reflow profile. Figure 10.A4 also shows the evaporation behavior of the solvent in flux, the flow behavior of the melted flux during preheating, and the solder flow behavior during main heating.

The surprising discovery in the observation of the fillet formation was an existence of a series of processes showing that melt solder wets on two electrodes of land and part electrode as follows. At first, the solder wets the land, then extends, and, moreover, flows again. Observing the detail of fillet formation more carefully, the melt solder flows while balancing the interfacial tension at the boundary of electrodes after two electrodes of land and part

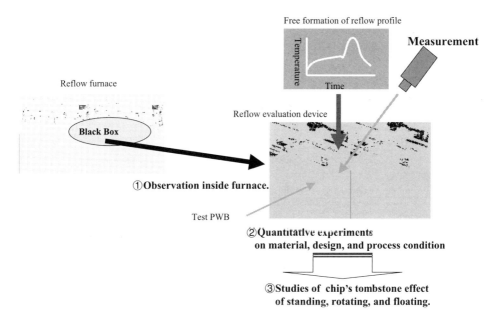

Figure 10.A2 Research opening a black box of reflow.

Table 10.A2 The Representative Defectives in Element Processes

Factor	Defect attitude	Process-originating defects		
		Print	Mount	Reflow
▪ Man factor	Between lots	Blot bridge	Part mistake	
Variation	variation		Direction error	
			Parts none	
▪ Machine factor	Entire line		Crushing too much	Wicking
Condition	Time passing		Part gap	Small fillet
Change				Excess solder
▪ Manner factor	Model commonness	Graze		Partial wetting
Amount of solder	Model commonness	Bridge		Floating
Supply position	Model commonness	Lack		Standing
Reflow profile	Model commonness			θ gaps
	Part commonness			
Nonwetting				Dewetting
				Nonwetting
▪ Material factor	Entire line	Slump		
	Time passing	Graze		

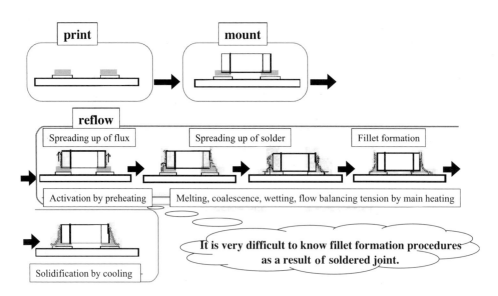

Figure 10.A3 Behavior of solder in fillet formation.

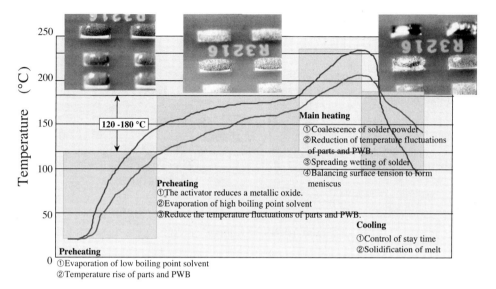

Figure 10.A4 Relation of behavior of solder and reflow profile.

electrode wetted. Finally, the meniscus of the solder is formed with the action of the surface tension of the solder at the cavity surrounded by two electrodes. The meniscus is an origin of fillet shape.

10.A1.4 Cancellation of Black Box by 3M Integration

To stabilize soldering, measurement methods of characteristics used to measure some procedures important to form fillet were developed. A reflow evaluation device of Fig. 10.A5 has been developed. The parameters that stabilize the soldering elements procedure were experimented on. "Prescription of the solder paste," "land design," "heating condition," etc., were woven to a test engineering group (TEG). Quantitative measurement was done as much as possible.

As a result, the stabilization of the reflow—that "prescription of the cream solder," "process condition of the reflow," and "substrate design" were mutually related to a triskelion—was clarified. In other words, the three Ms of material (solder), manner (part selection and part layout design), and machine (process condition) consist an engineering system (3M engineering system). If the 3M relation is understood, the defect can be cancelled. (Here, man is management related to the person, like the work standard and an inspection standard.)

Conventional SMT utilizing equipment   **Technological evolution that 3Ms integrate**

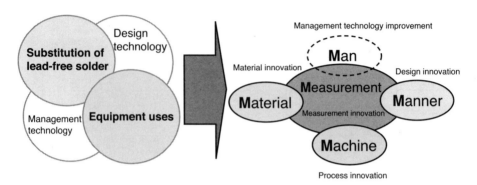

Figure 10.A5   Evolution of SMT.

Figure 10.A5 shows evolution of SMT in the future by comparing a conventional SMT and SMT based on 3M engineering system in the future. Current SMT was "mounting technology that utilizes equipment well" with highly miniaturized part using precise mounting equipment. Hereafter, it should be technological evolution based on integrated 3Ms. It is exactly the case for the change to lead-free solder. The defect might happen frequently, and a fear of the reversion of the constitution of the production came out when there was lack in the balance of 3Ms.

The indications for the improvements learned from 3M engineering theory are as follows:

1. It becomes easy to improve using relations of 3Ms (3M engineering theory).
2. Measurement of 3Ms is necessary for crowding in the process.
3. In the viewpoint of the inspection, the defect is a result of incompatibility in the 3Ms.
4. The effect is large in the order of material > manner > machine according to the substitution rule of IE.

(Man is a factor with a large defective uncertainty because the person varies with the machine factor change. Man's origin of defect is described in detail in another paragraph.)

10.A1.5   Measurement Management of Reflow That Does Not Make Black Boxes

The author made the reflow technology that did not make the black box by using 3M engineering theory. It was necessary to manage reflow easily by simple control using fewer variables such as: solder selection, part selection, and heating condition on the base of 3M engineering theory. With the generality of manufacturing, the material and the process out of three variables had to be made constant. Since it was examined, it was noticed that the production of SMT needs to reflow into a variety of models at the same heating condition. When the line is used for exclusive model use, the relation between the design and the process varies. In other words, the process condition changes whenever the design changes, and there is finally a possibility of turning back to the black box. The biggest factor to influence the process capability of SMT is 3Ms of reflow.

It returns to the custom of production of past specific lines to a model where 3M relation is disregarded, and manufacturing changes the process condition into the design. A problem that is bigger than if it is producible or not occurs as follows:

1. Past manufacturing results are not useful for startup of the new product.
2. Early evaluations of new parts will not be done.
3. Judgment of what design into which process model can be thrown is not understood.

Thus, the constitution is changing from time to time.

Then, the ideal way of the reflow that does not make the black box and the outline of the measurement management are shown in Fig. 10.A6. Figure 10.A6 also shows that reflow profile originates from the thermal capacities of parts, which we want installed in one board, layouts, and heating conditions, and shows that a reflow profile is a result of the origins.

Because the reflow method is used to operate a variety of models at a same heating condition (model-to-model mixing operation), the heating conditions are fixed. Two Ms of machine (heating condition) and material (solder) are constant among 3M simultaneous equations. Then, only manner (part selection) is a variable. Therefore, the reflow profile is predictable at once when the design is decided on using 3M theory. In other words, the reflow temperature can be presumed beforehand at the stage of the PWB design. The model-to-model mixing production is controlled in the range of upper and lower limits of the peak temperature, and in deciding on the heating condition of putting the peak temperatures of all parts in the range.

The author has not understood where the lead-free soldering was started either. In such a situation, the abovementioned 3M equations were not hit. The reflow profile should be measured at a correct position based on the principle of heating. Two profiles of red and blue show bound pairs of all parts recorded in a board. The profile of the upper bound makes it correspond to heatproof temperatures of parts, and the profile of the lower bound makes it

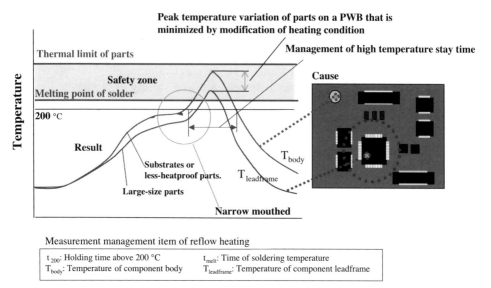

Peak temperature variation of parts on a PWB that is minimized by modification of heating condition

Measurement management item of reflow heating

t_{200}: Holding time above 200 °C	t_{melt}: Time of soldering temperature
T_{body}: Temperature of component body	$T_{leadframe}$: Temperature of component leadframe

Figure 10.A6   Measurement management of reflow that does not make a black box.

correspond to the melting point. The profile takes a high or low position between bound pairs according to parts on the board.

A correct way to measure the profile is to measure the temperature of the terminal parts where thermal capacity is large and soldering is difficult, and the temperature of bodies of parts that easily increase in temperature. Measured terminal temperature is compared to melting points of solder, and measured body temperature is compared to heatproof temperature limit.

The most difficult judgment in the reflow using lead-free solder is whether reflows using thermally weak parts are possible. The stay time over 200°C often becomes a restriction of the reflow than the difference of the peak temperatures. Therefore, it is necessary to measure not only the temperature measurement but also the stay time of this overheating. There are a lot of parts that provide for the stay time of 200°C as an index, which shows heatproofing of parts.

It is necessary to measure T_{body} (part body temperature), $T_{termination}$ (terminal temperature), $t_{200°C}$ (stay time over 200°C), and t_{melt} (soldering time) for the management of the reflow profile in lead-free soldering age [4]. The temperature can be easily managed without considering the complex principles of 3Ms because the temperature profile is simply measured like this.

10.A2 THE TRINITY OF DESIGN, PROCESS, AND INSPECTION FOR DEFECTIVE EXTERMINATION

The reliability of soldered joint can be controlled according to the solder selection, the fillet design, and the manufacturing condition. When lead-free solder was selected, the author made the sample by using the reliable fillet design and tested the reliability. The reason is that the design of the fillet shape strongly influences the reliability. Moreover, the land design that made it easy to inspect beforehand was used in the inspection. Here, the defective extermination and the easy inspection—especially from the design side on the base of the integration of the design, the process, and the inspection—are described.

10.A2.1 Building the Demand Quality of Markets in the Fillet Shape

It has been understood that the rheology of melt solder controls the fillet formation in the procedure of the reflow, and that fillet shape can be formed by using 3M controls as described at the preceding clause. A variety of fillet designs are used from a recent thin-and-small fillet to a without-fillet joint, along with advancing high mounting density. What shape in fillet is suitable for judging from the reliability viewpoint? Industrial reliability is represented by the thermal cycle endurance from its use in the market.

The effect on the reliability of the fillet shape was examined by a thermal cycle test ($-50\,°C$ to $+125\,°C$), making collaborations of various fillet shapes by using the chip components. The variations of the fillet shape and the results in measuring the shearing strength of the samples with the variation in shape provided difficulties in the reliability judgments. Then, the thermally induced stresses that occurred in the joints of the chip components were analyzed by using CAE. The result is shown in Fig. 10.A7. It is shown in red and blue in the figure in the order of thermal stress strength. When thin fillets of joints are compared with thick fillets of joints, it is understood that the size and the position of the stresses that occur in both are different in Fig. 10.A7. A thin fillet of joint has the tendency on which the thermal stress does not concentrate. On the other hand, the stress strongly concentrates on the upper part of the fillet as for a thick fillet of joint. It is understood that a thin fillet of joint is excellent in the thermal cycle endurance from this. As for a thick fillet of joint, a large stress concentrates on the interface of the part electrode and the solder. In an actual reliability test, the crack enters from the upper part of the fillet of joint. Once the crack starts from the surface, thermal stress is transmitted along the interface of the part and the solder, and the crack finally spreads to the whole fracture. It can be understood that a thin fillet of joint develops, on which the stress is not easily concentrated from both sides of the experiment,

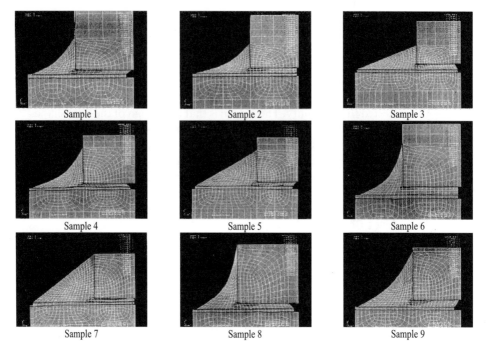

Sample 1   Sample 2   Sample 3

Sample 4   Sample 5   Sample 6

Sample 7   Sample 8   Sample 9

Figure 10.A7   Fillet shape design of chip parts for long life. CAE analysis of thermal stress that occurs in a variety of fillet shapes for chip parts under thermal cycle test with the cycle of $-50\,^{\circ}\mathrm{C} + 125\,^{\circ}\mathrm{C}$.

and then the analysis becomes highly reliable. The author adopted a thin fillet of joint for the fillet design of industrial use, which values the thermal cycle endurance according to the findings for this fillet design to control the degree of the thermal stress [2].

Next, the same experiment and analysis were performed on QFP. Figure 10.A8 is the result of the same analysis of the chip components of QFP. In this figure, the thermal stress concentrates on backfillets of leads. Breaking actually starts here. In the collaboration of the lead structure such as QFP, this finding and the thermal stress analysis (that the thermal cycle test is known in general) correspond well. As the tensile test used in experiment and the thermal stress analysis assume the same breaking position, the correlation of these two evaluation results mutually corresponds.

However, there are many cases where a selection of the solder with large mechanical strength or large fillet shape is designed contrary to the idea of the abovementioned. In other words, the ideal theory of the fillet design is not decided yet. In this background, the facts (a) that shearing strength as a char-

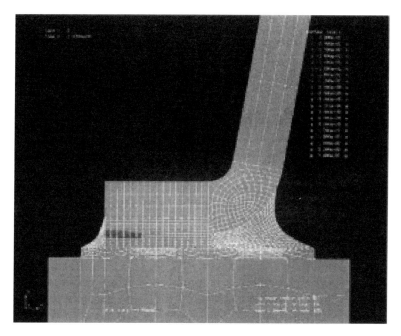

Figure 10.A8   Fillet shape design of QFP for long life. CAE analysis of thermal stress that occurs in the fillet for QFP under thermal cycle test with the cycle of −50°C to +125°C.

acteristic of the reliability evaluations for the chip components is generally used in measurements and (b) that a thicker fillet gives larger mechanical strength are related. Higher mechanical strength does not always give stronger thermal cycle endurance from the abovementioned result. Moreover, on the manufacturing side, a thick fillet of joint needs a larger amount of the solder supply, which easily occurs in defects such as bridging, solder ball, and θ gap.

The relations between the fillet shape and the thermal cycle endurance are mainly summarized as follows:

1. The thermal cycle endurance can be made strong in the fillet shape controlling the solder supply of the amount and the position.
2. It is necessary to choose the fillet shape so that reliability is strong enough in each part to evaluate the reliability of soldered joint. The effect of the shape selection is usually superior to the effect of the selection of solder.
3. A lead-free solder is generally more highly reliable than Sn–Pb solder. If the suited fillet designs are not used for individual terminal

shapes of the part structures of the terminal shape in a variety of parts, the superior reliability of lead-free solder cannot be utilized.

10.A2.2 Design That Values Reliability and Correspondence to Inspection for Various Fillets of Soldered Joints

The designs of the fillets of joints are indeed variedly used. The variety of the fillets cause difficulties in solder inspection. The reasons as to why the fillet design is various are: the high mounting density of electronic equipment requires small fillet, and there are various necessary levels of reliabilities according to the usages of equipment. In fact, the reliability levels needed are different in household use and industrial use.

In automobile use, the amount of solder has the tendency to become larger because it needs vibration-proof performance. Thinking of the vibration proofing, only the idea from the PWB unit or the fillet of joint to design antivibration is not suitable. First of all, let us think that it is important to fix PWBs to the assembly case at specific positions considering that the structure assembled with the PWB and the case are in one vibration system, and to reduce the vibration of the entire structure. Moreover, when the mass of parts installed on the PWB is large, it is preferable to bond. It is not a correct to use solders to apply structural force or repetition load to solder joints. The reason is that solders have the creep characteristics. The fillet design for miniaturized SMT could be shared with industrial use when standing in assumption to give antivibration performance to the case design. The without-fillet joint generally used in cellular phones must be avoided in industrial use or automobile use. The reason is that fracture easily starts at the interface of the part and the solder if there is no fillet of joint.

The heat cycle endurance can be improved by switching Sn–Pb eutectic solder to lead-free solder such as Sn–Ag–Cu. However, the level of improvement does not increase by as much as 10%. Moreover, the characteristic of solder is not uniformly effective in all parts either. For example, it hardly changes after the heat cycle test because the aluminum electrolytic capacitor has a flexible terminal with round pins. Conversely, the large chip component easily gives influence on reliability because of its hard structure and the large thermal expansion. Therefore, it is necessary to design fillets of joints properly in various parts to raise the reliabilities. There is a formula to improve joint reliability by accumulating the designs as a library and by using the design.

10.A2.3 "Land–Mask Combination Design" Integrating Manufacturing and Design

Designing a fillet properly in various parts was described and the utilization enabled the reliability improvement in Sec. A2.2. Here, we describe the procedure on how to design fillets properly.

First of all, what do you think is the proper fillet shape? The conclusion is a fillet design that improves all of "the first reliability, the second manufacturing, and the third inspection." Figure 10.A9 shows the method of designing the fillet of soldered joint. First of all, the "amount of the solder supply" is calculated. Then the "size of a mask open mouth" from the long life fillet shape of joint and the size of the electrode of parts is calculated. The basis of the print is to transcribe a necessary amount of solder. Even if the fillet shape to bring high longevity is complex, the amount of solder supply can be easily calculated by dividing into the plural of F1 and F2, etc.

Next, "relative positions of the land–mask parts" are decided from the evaluation experiment on manufacturability. The opening mouth area of the mask is given by a ratio of necessary volume of solder/mask thickness.

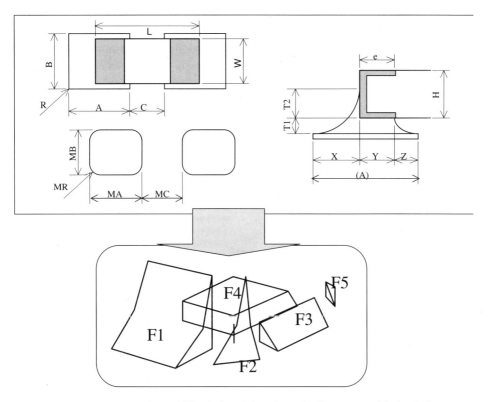

Figure 10.A9 Procedure of fillet design. Manufacturing integrates with the design to use "land–mask combination design."

The library for lead-free solder is not necessary based on the idea of the fillet design of the reliability value described at Sec. 3.2. It is very possible to share the library for Sn–Pb eutectic solder and lead-free solder.

10.A2.4 Fillet Design and Automatic Inspection

When manufacturing and design have been integrated by "land–mask combination design," the fillet shape of joint is made by 3M control. "Inspection standard" is decided from the quality design and the process design, and the quality is inspected by the inspection standard.

Figure 10.A10 shows the procedure for setting the inspection standard and the verification condition. "Limit sample" of the quality item should be provided when the visual inspection is done during production. Keep the reproducible judgment by teaching to the worker frequently. The limit sample is defined in the fillet shape where minimum level of necessary reliability (limiting quality) is secured. The shape is defined by the characteristics that can be measured (e.g., "height of wetting"). A 3D measuring instrument is used in the measurement.

Paying attention to the difference of the shapes of defective goods, the limit samples, and the quality items, the feature shape of the limit sample is selected. The characteristic used for the automatic inspection is a substitute characteristic different from the measurement value of the limit sample. In other words, the measurement value of the limit sample of the visual inspection is rewritten in a new standard value of a substitute characteristic in the automatic inspection. The conversion of the inspection characteristic of automatic inspection to the measurement characteristic of the limit sample is ex-

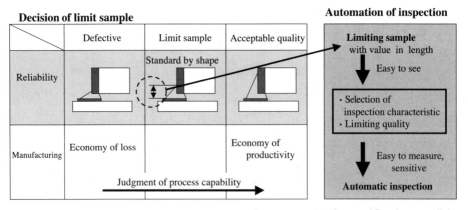

Setting verification conditions

Figure 10.A10   Inspection standard and the procedure.

plained by taking an example of the solder inspection machine using the color highlight method. The RGB color light source is converted into the color image by the reflection on the solder surface, and the height of wetting (limit sample measurement value) is detected as "change in RGB pattern" (inspection characteristic) corresponding to the height of wetting.

In the automatic inspection, there is an advantage whereby sensitivity improves more than the actual inspections, by using a substitute characteristic. The automatic inspection achieves another effect of labor reduction, and the reproducibility increases during inspection (prevention of overlooking).

10.A3 USE OF INSPECTION RESULT TO IMPROVEMENT

10.A3.1 Process Capability Analysis (Macroanalysis) and Defect Origin Analysis (Microanalysis)

Even if the inspection machine is adopted, the process capability of SMT cannot be improved. It is necessary to improve the design and the process to increase the rate of the quality item.

There is a role of "defects do not flow out to the next process" in the inspection machine. Besides, there is another role for "giving the directions for improvement from a defective attitude, etc." An appropriate, effective improvement becomes possible by positively using the inspection result.

How to use the inspection is roughly bundled as follows:

1. The process capability is understood from the generating state of defectives (macroanalysis).
2. Defective originates are specified from findings of the productive engineering and the generating state of defectives (microanalysis).

Figure 10.A11 shows the use of inspection result. Inspection result is used to estimate the process capability from the generating state of defectives. As for the macroanalysis, the defect of the entire line is totaled, and the process capability from defective total and the breakdown is estimated. Making the Pareto chart, defectives with high incidence are cleared, and the range of the improvement is confirmed. Whether it suddenly improves from the high ranks of the number of generation, or improves from ease of performance or cost to improve is considerably judged by the microanalysis described later.

10.A1.2 Grasp of Improvement Points and Analysis of Improvement Effect

Figure 10.A12 shows the view of data to analyze the inspection result when a production engineer improves defects. One looks after defective data to know "when and where does the defect go out" based on the engineering findings.

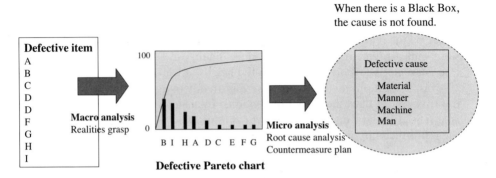

Figure 10.A11   Use of inspection result. Grasp and defective origin analysis of process capability.

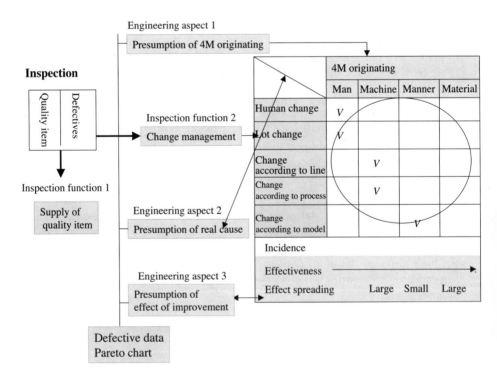

Figure 10.A12   Engineering aspect in inspection result analysis.

In the emphasis on high-ranking defect of the Pareto ranking as a microanalysis, "state that the defect occurs" is confirmed by looking at the inspection result from a variety of change elements according to lines, lots, processes, models, and parts.

On the other hand, defective origin is presumed fully using 4M engineering aspects. Concretely, the system of defective representatives according to process finding "typically occurs because of origin of the material, the board design, the process condition, and the person who is set up beforehand. The defective origins (where and when) are analogized because actual defective generations are collated with defective representative origins." Table 10.A3 shows examples of representative defects according to main processes.

Next, integrating the aspect from the change element and the aspect from 4M, the real cause (why) of defects is presumed. There must be possibilities of four originates because the defects relate to 4Ms.

The process for effective improvement is shown in Fig. 10.A13. Efficient improvement is planned by presuming the improvement level and extension of the effect using the findings of IE and Pareto analyses. The countermeasure and the expenditure from total defectives according to the presumed origin are presumed. The improvement cost is calculated. The calculation of effect/cost from results of the improvement according to origins provides economical plan for improvement.

Thus, the object of improvement can be effectively narrowed by microanalysis, and an effective improvement becomes possible by using inspection result and accumulated improvement skill.

Table 10.A3 State Watch Measurement in Process

Process	Purpose of measurement	Object of measurement	Measurement method
Print	Print defect prevention	Solder print factors	Squeegee pressure, parallel level, viscosity, and tackiness
Mount	Excessive gap prevention To mount parts	Mounted position	Image processing
Reflow	Maintenance of heating condition Easy to reflow design	Temperature and time of profile	Measurement of heating energy Change speed of profile

The above-mentioned measurements are currently done by off-line measurement.

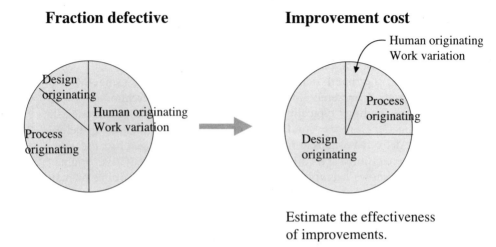

Estimate the effectiveness
of improvements.

Figure 10.A13   How to find effective improvement.

10.A3.3   How to Proceed Effective Improvement

Figure 10.A14 shows how to advance an effective improvement. Start from man according to the formula of IE, and improve 4M in the order of man, machine/material, and manner.

The points of improvement are as follows:

1. Observe the manufacturing standard completely, and the number of change points must be one for one time.
2. The improvement is on the order of observation, process change including solder, and design change as a formula.
3. The standardizations of the design and the process become assumptions.
4. The material that suits the process characteristic should be selected beforehand.
5. Model-by-model mixing flow of the reflow becomes a big productivity improvement.
6. The setting up of improvement target (to use the inspection result and the change management of the improvement using the inspection machine) is necessary.

Improvement to Man (Case 1)   The person's quality and mistakes relate to this defective. There are defects due to the work change and mistake in the

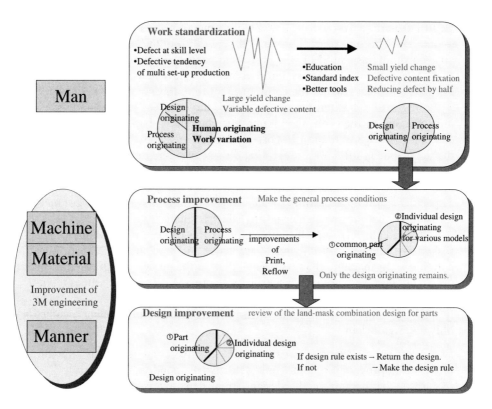

Figure 10.A14 How to advance effective improvements.

setup, and defects due to process change that lacks skill causes. Moreover, it is a countermeasure for the process maintenance.

When the person does not settle, when work is not steady, and when the setup of the machine is not steady, the yield violently changes and the defective content also changes day by day. Then, teaching the person, standardizing work, and stabilizing the machine are performed. When the work points fix, the defect roughly reduces by half. As for the remaining defect, the process origin and the design origin become the majority. When unstable work stops, improvement by 3M engineering becomes significant. It is a formula of improvement to change one parameter at a time.

Even if improved under immature work and unstable work, the shake of the yield is large and the effect degree could not be read as it is hidden in the width of the shake.

Improvement to Machine (Case 2)   In general, process improvement and maintenance become objects. Because the crowding of the quality of conformance becomes basic in SMT, the process improvement is important.

The feature of process improvement is the extension of the effect. Because the effect reaches the entire product that flows in the line connected with the cereal, the extension of the effect is large. In SMT, do an individual improvement of the print, the mount, and the reflow. Especially, the reflow can achieve large reduction in costs and extremely high productivity (production capacity for line) by operating the line using model-by-model mixing flow. The reason is that the effect of the improvement reaches even other lines to unite the heating conditions. Oppositely, the line that operates under a specific condition according to the model becomes an individual countermeasure of the improvement, and the effect is small.

Maintenance is equally important to the process improvement. The reason is that, with an equivalent influence to the change in the process condition, the machine changes though neither the process condition nor the design change. The mounter is a process where SMT especially values maintenance among three processes.

When the process improvement advances, the remaining defect becomes only a design origin.

Improvement to Manner (Case 3)   If defective content is seen after the process is improved, it becomes only defective in an individual substrate. The feature in the attitude of the design origin defective is the specific board. Common defects to specific parts across two or more boards exist when a defective board is seen as well. In both cases, correction of each board is needed, and improvements cost us a lot.

The cause of the design-originating defect is analyzed as follows:

1. Design error that is not the defense of design standard (when there is a design standard)
2. Evaluation shortages of new parts beforehand (when there is no design standard).

The real cause of design-originating defects is either of the two abovementioned cases.

The design is returned in case 1, and trying to make it in case 2 becomes countermeasure. First of all, when there is no design standard, it is difficult to specify a defective cause because defective generation is too complex.

In the design improvement, it is necessary to execute the improvement by thinking about the economy of the investment and the labor.

Improvement of Material (Case 4)   The material is called a measure that provides the maximum effect if it makes it good and also a measure that provides the maximum side effect if it makes it bad because of the idea of 4M substitution. It is preferable to adopt the solder with a good result beforehand in the improvement because a large effect can be expected because 3M engineering is a triskelion, as mentioned before. It is necessary to verify the influence in the print and the reflow condition beforehand when the material has been changed after the process improvement and the design had been improved. Therefore, the author provides a necessary characteristic of the solder paste to each process function of the print, the mount, and the reflow, as shown in Sec. 2.2, Therefore, the specification change is not needed as long as the machine condition is not greatly changed.

10.A4 CORRESPONDENCE TO THE CHANGES OF PRODUCTION FORMS

Not only the technological change of lead-free soldering but also the market change of supply chain management in e-business give influence to the production form evolution of SMT. Conventional production of SMT has achieved a very low cost of production with high-speed productivity. However, the problem of producing too much has occurred, on the other hand. In mass production, it only has to have a specific line for each model. However, because the age of SCM has come, it is necessary to produce various products of different models little by little by the necessity on one line (a modicum production way of model-by-model mixing flow). Figure 10.A15 shows the change in the production form in the future and the change in the inspection. The production form of SMT starts to change form from the specific SMT line according to each model, to model-by-model mixing flow SMT, and, furthermore, to SMT synchronization with assembly. It is forecast that the inspection will expand the roles of inspection from a defective inspection to the monitoring of the nondefective processing and, furthermore, to the monitoring of productivity improvement.

   The inspection machine is evolving to include not only the inspection function, but also the effect judgment of the improvement and the function of the productivity monitor. The future system that integrates off-line engineering, off-line measurement, and in-line inspections and measurements is shown in Fig. 10A.16. This figure is the one in which the improvement made by the author was described in the system chart. This system will effectively work to keep evolving in the future.

   In the production that synchronizes SMT and assembly, a movement to increase the elasticity degree of production with small machines of the cell type is active. The movement of soldering improvement, from utilizing only machine performance to changing the soldering system itself, has taken place.

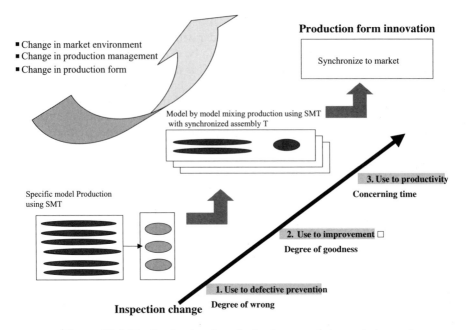

■ Change in market environment
■ Change in production management
■ Change in production form

Production form innovation

Synchronize to market

Model by model mixing production using SMT
with synchronized assembly T

Specific model Production
using SMT

3. Use to productivity
Concerning time

2. Use to improvement □
Degree of goodness

1. Use to defective prevention
Degree of wrong

Inspection change

Figure 10.A15   Production form in the future and change in inspection.

10.A5 "CRITERIA OF INSPECTION" FOR THE EVOLUTION OF PRODUCTION

The evolution of the production in the future will be diversified. One might be densified and the other one flexible. It is a big role for the inspection machine to clarify the effect of the production evolution. Table 10.A4 shows the criteria of the inspection for the evolution of production. The evolution criteria of soldering inspection are regulated to four stages of: "reduction of the defect," "production design standard," "technology innovation," and "making of the constitution":

1. Stabilize the routine production (reduction of the defect)
2. Attest to an existing good product design (production design standard)
3. Effect of verification of new parts and new processes (technology innovation)
4. Production not broken by changes (making of the constitution).

In the future, among the four steps shown in the table, the third step and the fourth step might become the future image of the inspection. To keep evolving without allowing the constitution to turn back, the idea and 3M

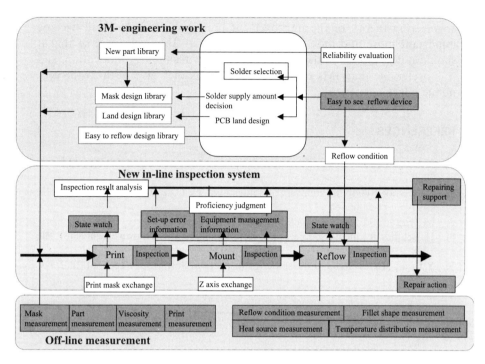

Figure 10.A16 Measurement and inspection to use jisso innovation.

Table 10.A4 Criteria of Inspection for Evolution of Production

- To stabilize the routine production
  Verification that processes of manufacturing were appropriately done
    Defective outflow prevention to next processing
    Discovery of change in process (early stage management)
    Measurement of the maturity level of the process
- To turn the design of good production results to the upstream
    A good design is shown from the quality item data
    Find the improvement points from defective data
    Predict the effect of the improvement
- To manage the early stage instability confirmation of a new design/technology
  Management in a new part, a new design, and a new process
    A new part is verified that there is no problem in manufacturing
    A new process is verified that there is no problem in manufacturing
- To make the robust constitution of "never brake" in the process
    Compulsion by system
    Automatic generation of standards
    Management of efficiency operation and economical information
    Management that integrates SMT and assembly

engineering theory of the improvement previously described will become important more and more. An individual M of 3M is changed so that it should not give birth to the black box with new technology innovation, and it is preferable that 3Ms evolve sequentially as they give birth to the economic effect.

REFERENCES

1. M Hirano. Practical use of lead-free solder and R&D scenario. JIEP 3(5):439–444, 2000.
2. H Bessho. Reliability analysis of solder joint by CAE. Omron Tech 37(1):20, 1997.
3. M Hirano. Surf Mount Technol 6(7):18–22, 1996.
4. H Bessho. Reducing temperature variation in reflow. Omron Tech 38(3):240, 1998.
5. Y Fujita. Latest trend of the soldering inspection. Surf Mount Technol 18(2):66–71, 2002.

11
Future of Lead-Free Soldering

Katsuaki Suganuma
Osaka University, Osaka, Japan

11.1 SUMMARY OF CURRENT LEAD-FREE SOLDERING TECHNOLOGY AND BEYOND SOLDERING

In the previous chapters, we discussed the many aspects of this technology from the basic science to technology with legislative movements. The standard material as lead-free solder is, needless to say, Sn–Ag–Cu. Table 1.2 already showed the brief summary of composition, feature, some notes in application, etc., for lead-free solders in the market. For Sn–Ag–Cu alloy, there are slight differences in their compositions in Japan, the United States, and Europe. However, these compositions can be accepted and are available all over the world. These solders are most stable and possess good compatibility with the present technologies, e.g., components and their various platings, and high reliability. Especially, as their mechanical property is good, it is possible to extend the life of equipment by applying optimum mounting design and process control. As regards surface mounting, the property of lead-free pastes almost reaches the same level of the conventional Sn–Pb solder; in fact, there is no problem in productivity on many electronic products.

On the other hand, due to increase in the melting point of solder, it is necessary to give more careful attention to temperature control. Figure 11.1 compares the typical lead-free solders as a function of melting temperature [1]. The demand for solder having lower mounting temperature is extremely high

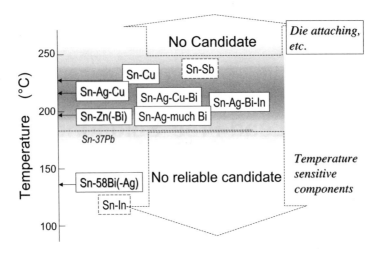

Figure 11.1 Melting temperature of lead-free solders and technology open pocket. (From Ref. 1.)

in the fields of products, which are sensitive to temperature such as chip sets or liquid crystal display (LCD) panel of note PCs, organic electron luminescence (EL), light-emitting diode (LED), quartz, etc., or large part or mounting on a board which has big a difference in temperature, or a thin-type multilayer board aiming weight/volume reduction, etc. Besides the above, the other important advantage of lower-temperature mounting is desirable to environmental protection by energy conservation and by CO_2 reduction. Even if lead-free product is developed, if process temperature is extremely increased, it will be harmful to environmental protection, which is not true for lead-free soldering, unluckily. It is desirable to lower-mounting temperature at the range of the same level or even lower than that of the conventional Sn–Pb eutectic solder.

As already mentioned, some of the solders, such as Sn–Zn, Sn–Ag–In, and Sn–Bi, enable this low-temperature soldering. Typically, Sn–8Zn–3Bi can be used at 210°C, Sn–3Ag–8In–0.5Bi can be used at 220°C, and Sn–58Bi can be used even lower than 200°C. One needs to understand the key factors and the limitation in the applications of these solders. In case of Sn–Zn-type solder, by remarkable improvement of paste, mounting in air atmosphere becomes possible, and application examples have been expanding to note personal computers (PCs), televisions (TVs), liquid crystal displays, printers for business use, batteries, personal digital assistants (PDAs), etc. However, it is required of Sn–Zn solder to have know-how for securing both productivity and reliability, and it is not a level which anybody can handle with satisfaction. Thus, development of further stable material is required. In case of Sn–Ag–In solder, the melting point can be reduced to 206°C at 8% In, and it is expected as the replaceable material with Sn–Zn one. As In is of higher value and a rarer metal than Ag, the application of Sn–Ag–In-type solder is limited to special applications and we must establish a suitable recycling system for this solder. Needless to say, Sn–In eutectic alloy cannot be used for many products. Sn–Bi-type solder has lower acceptable heat resisting temperature owing to very low melting point and is brittle and weak against impact. The temperature limitation may lie around 80°C. Brittleness of this alloy also limits its applications. Thus, we have to say that there is no adequate interconnection material, which makes lower-temperature mounting possible at present.

On the other hand, let us review higher temperature ranges. There were two significant viewpoints to solder for high-temperature mounting. One of them is, needless to say, the material which enables to bond in a range of temperature for over 250–300°C. This is a high melting point solder which is applied to inner bonding of semiconductors or step soldering (material which is applied to primary mounting of module and will not dissolve at a reflow step of the secondary step). For this purpose, solder having Pb for almost 90% has been applied. However, at present, there is no adequate replaceable material. Hence, in the European Union (EU) directive, this item is listed in the exceptional items. The other viewpoint for high melting point solder is not mounting temperature, but heat-resisting temperature. In other words, there is no proper solder which can be applied at around 150°C as the normal usage temperature. In an electronic field, one may feel that 150°C is the special high temperature. However, this temperature is the required condition for instruments for automotives, especially various control equipment, which are equipped in an engine room. Even at this time, there is no proper solder that is resistant to this high temperature, not even both lead-free solders and Sn–Pb solders. Both solder materials and the interfaces with electrodes degrade severely. Figure 11.2 shows the change in joint strength of Cu soldered with Sn–38Pb and with Sn–3.5Ag, subjected to heat exposure at 150°C. From this figure, we can see that the joint strength decreased gradually for both solders. This phenomenon is attributed to the coarsening of the microstructure of solder itself and the extreme growth of brittle intermetallic compound/voids due to severe reaction at interface. It can be said that it is not possible to prevent this phenomenon as far as we use Sn alloy basically. Thus, we have no established choice as high-temperature solder.

The list below shows the main concerns remaining to be completed for lead-free solder. Among them, one could list the fact that even eutectic solder has not yet been fully investigated, but in this unexplored region as well, trends have appeared to bring solutions that accompany development efforts

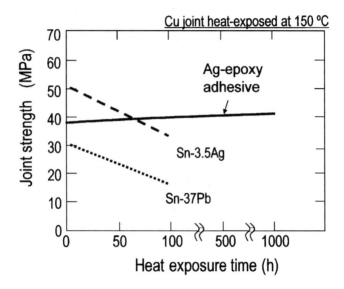

Figure 11.2 Tensile strength change of Cu joints with Sn–37Pb, Sn–3.5Ag, and Ag epoxy due to high-temperature exposure at 150°C. (From Ref. 2.)

for lead-free solder. Unsolved technical and scientific problems for lead-free solder as well as for leaded solders are as follows:

1. Clarification of lift-off phenomenon and establishment of suppression measures;
2. Establishment of lead-free plating technology and whisker countermeasures;
3. Lower soldering temperatures and process optimization;
4. High-temperature solder;
5. Low-temperature embrittlement (Sn pest);
6. Construction of a database of physical properties [solder, parts, printed circuit boards (PCBs)];
7. Establishment of reliability design technology;
8. Standardization of solder materials evaluation technology.

Sn–Pb solder has been easy and convenient to use and has been seen as the "grand champion of solders." However, this is not an accurate portrayal. The use of Sn–Pb solder has caused a number of failures and accidents. On the other hand, lead-free solder, represented by Sn–Ag–Cu, can certainly show improved reliability under suitably selected application conditions for mounting. Sn–Ag eutectic solder has clearly shown high reliability in its use to date.

Therefore, if we return to the starting point of lead-free solder mounting, we can immediately stop the continued release of lead into the environment and we can accumulate technological know-how. We must establish highly reliable lead-free solder-mounting technology at the earliest possible moment.

The standard solder of the future will not specifically be only Sn–Ag–Cu, but we must also ascertain the possibilities and range of suitability for such solder types as Sn–Zn, Sn–Bi, and Sn–Cu. To select correctly from among the possibilities which alloys and processes will become the final standards, we need to immediately construct a database; in order to realize such objectives, we need to organize strong cooperation among industry, government, and academia.

11.2 COMPETING INTERCONNECTION METHODS

Figure 11.3 exhibits the relationship between processing temperature and heat resistance temperature of solders and other competing interconnection methods. One should notice that there are other unexpectedly excellent joining methods besides soldering. Table 11.1 summarizes rough comparison of each feature of these technical selections that enable low-temperature mounting. As we stated above, in lead-free solder, Sn–Bi and Sn–In types

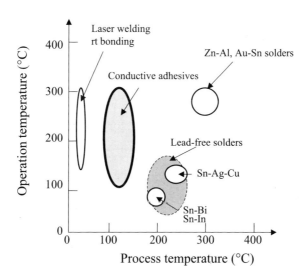

Figure 11.3 Various joining materials and their process and operation temperatures. (From Ref. 1.)

Table 11.1 Comparison of Various Low-Temperature Microjoining Methods

Methods	Feature	Pitch limit	Temperature limit
Conductive adhesives	All types of fine-pitch joining Compatibility with Sn plating	10 μm	300°C
Lead-free solders (Sn–Bi, Sn–In)	Poor reliability	130 μm	Below 100°C
Laser welding	Only metals Not for FPDs such as LCD, organic El Not for ceramics Limitation in shape and size	Laser depends on pins (a few mm); not applicable for area array type	Depends on metal
Room temperature bonding (pressure joining)	High cost Limitation in shape and size Need pressure	a few hundreds μm	Depends on combination

enable to apply low-temperature mounting. However, both alloys have many obstacles before practical applications. Of course, it is not possible to use solders containing much amounts of Pb or Cd that have been used for a long period. Laser welding and room temperature bonding in a solid-sate are attractive, but not suitable for practical usage in general purpose. Laser welding cannot be applied to fine-pitch mounting. The current one allows us to make bonding pitch in about millemeters order. In contrast, room temperature bonding with the aid of nonconductive adhesives, sometimes with a little heat, can be applied to the interconnection of fine-pitch array such as bear chip mounting, while it cannot be used for the conventional mounting of general components.

On the other hand, there is another selection as electric conductive adhesive besides solder. This enables us to apply low-temperature mounting at 100–150°C for general surface mount devices and has high heat resistance which can resist even at 300°C for few hours. Figure 11.3 also shows the heat resistance of Ag epoxy-type adhesive as typical conductive adhesive as well as heat resistance of solders. Although solders loose bonding strength gradually, bonding strength of the conductive adhesive does not change even after 1000 hr; under exposure at 300°C for 1 hr, it still maintained about 70% of initial bonding strength. Nevertheless, making conductive adhesives, possessing such excellent features, as alternatives to solders brings several problems. For instance, the compatibility with Sn plating is one of them. Degradation of interface between Sn plating and Ag epoxy adhesive occurs at high temperature of 150°C and also under a humid atmosphere. The former is caused by unidirectional diffusion of Sn to Ag layer [2] and the latter is caused by galvanic corrosion reaction. The recent research work improved those compatibility problems of conductive adhesive and several improvement methods have been proposed [3]. Thus, the competing interconnection methods have their own benefit and drawbacks. We need further investigation and accumulation of technical knowledge on them as well as the further advancement in lead-free soldering.

11.3 WORLDWIDE STRATEGY OF ELECTRONIC PACKAGING

Recently, electronic production by lead-free solder becomes worldwide tendency and one can find many practical products already in the market. In a practical mass production by using lead-free solders, Japan has been leading in the world. However, WEEE/RoHS has already stipulated and the world does not only look at the success of practice in Japan. In fact, in Europe, which is based on WEEE/RoHS, the academic project, which is in cooperation of around 50 universities, and the IMECAT project, which composes mainly of industrial field and makes Japan to its target, have been applied widely and have been in a situation of pursuit preceding Japan. The NEMI and many companies in the United States become quite an active movement to lead-free soldering. In Asian countries, which have extremely low production cost, many branch factories of Japan, the EU, and U.S. companies have begun the production of lead-free equipment. From the other viewpoints, the competitors, which Japan, the U.S., and European countries will face, are not only themselves but also Asian countries of rapid advancement. Under the serious situation that production in countries has become difficult gradually, it is the most important subject at present to find the right point of view for domestic production in the future.

   One of the solutions in this matter is mounting field, which has higher reliability in fine technology. The fact for the shift of factories from Japan, the US, and Europe to Asian countries is that of soldering lines to overseas. It is required to return soldering lines to their own countries. Otherwise, it is hard to prevent this tendency. "What can they do for their countries to overcome this cost competition?" This is big subject for them. There is a chance to solve this problem for them, e.g., ultra high-density mounting, which has been and will be used in high-performance cellular phones and information home electric appliances. Figure 11.4 shows the transition and future expectation for the next decade of tendency of fine-pitch for flip-chip technology [4]. Devices have became smaller and smaller as the promotion of intensive integration. From a tiny device, a lot of bumps must be connected into circuits of a PWB correctly. At present, the practice of 130-μm pitch is possible; in future 10 years, the development of half size of the present pitch is expected. If the general target is 60 μm, one must make his target to the half, e.g., 30 μm, which cannot be achieved by soldering anymore.

One of the keys for opening door is new technology, which was not settled yet. Nowadays, we have a nanotechnology; for instance, by using the "nanopaste" with metallic nanoparticles, it becomes possible to make a circuit of resistance lower than that of solder. In both the low-temperature mounting and the heat-resisting mounting, electric conductive adhesive is superior to lead-free solder and may compensate completely blank in technology. One can expect big advancement in future development in this technology area. Of course, nanotechnology, which neglects economic factor, is useless. In the future 10 years, it is required for mounting to overcome many barriers for micron order. Many people believe that nanotechnology is an essential factor for new technology. However, one has to understand that it is not possible to produce any product by relying only on nanotechnology. However, recent development speed of materials is remarkable. There is a possibility of development on epoch-making nanotechnology in the future;

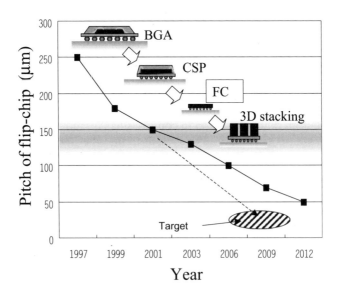

Figure 11.4 Fine-pitch roadmap for a flip-chip technology. (From Ref. 4.)

none exists at present. In the field of mounting, "real applicable" nano-technology, which has both economic advantages and fine properties, is really expected.

The other viewpoint suggests that the new market lies in front of us at higher-temperature usage. There are two meanings for "high temperature." One of them is the application of heat resistance beyond $150\,^{\circ}C$ as mentioned earlier. The typical applications are automotive ones. The field of automotive applying fuel cell and hybrid system on which further high heat resistance and higher reliability are required will further expand as a new market of electronics. Not only engine control and audiovisual (AV) systems but also improvement of communication performance will accelerate the tendency of automotive applications to information technology. The other requirement is high temperature lead-free solders. High-temperature solder was exempted from the RoHS at this moment because of technological difficulty. Nevertheless, it makes the recycling system complex because much Pb still exists in components. Someday, the prohibition must be there. There have been several reports already on high-temperature solders such as Bi–Ag, Zn–Al(–Mg), Zn–Sn, and conductive adhesives. Much work is needed to establish such high-temperature applications of lead-free solders.

In the field of advance electronic packaging technology, it is required to envision the following:

$$\left.\begin{array}{l}\text{Elimination of hazardous substances}\\\text{Making disposal and recycling easy}\\\text{Extension of products life}\\\text{Future fine-manufacturing technology}\end{array}\right\} \longrightarrow \begin{array}{l}\text{Green Products}\\\text{Strategy}\end{array}$$

Of course, the best efforts for development in the industrial field is essential; however, it is also required to enforce backup study system of universities and institutes to support this effort. WEEE/RoHS has a complete system that checks on the degree of advancement for technology development always and reviews its content. Nanotechnology and environment-friendly electronic packaging technology are two keywords of electronic packaging in the next generation. Besides these, an extremely attractive fine function and an advance packaging technology are expected to give future businesses a chance of worldwide capacity. We hope that our R&D effort will reap fruitful results.

REFERENCES

1. K Suganuma. Soldering for Beginners. Tokyo: Kogyo-Chosakai, 2002.
2. M Yamashita, K Suganuma. Degradation mechanism of Ag-epoxy conductive

adhesive/Sn–Pb plating interface by heat-exposure. J Electron Mater 31(6):551–556, 2002.
3. M Yamashita, K Suganuma, M Komagata, Y Shirai. An improvement of conductive adhesives on high temperature endurance by using Ag–Sn alloy powder. First International IEEE Conference on Polymers and Adhesives in Electronics and Photonics (Polytronic 2001), Potsdam, 2001, pp 265–270.
4. Electronics Packaging Roadmap in 2010. Tokyo: Japan Institute of Packaging, 2001, pp 121–123.

Index